SET OUR CHILDREN FREE

Tony Caruso

Editing

Ruthann Gambino
John Miller

Set Our Children Free

Copyright © 2010 Tony Caruso

All rights reserved.

ISBN: 0615469841
ISBN-13: 9780615469843

Foreword

Every story and event in this book is true. Every student, teacher, and administrator mentioned is a real person. There are no "composite" people. Only their names have been changed for obvious reasons. Some of these stories will shock you, and some will entertain you. They had that same effect on me as I was experiencing them in real time. There were times when I was laughing on the outside and crying on the inside about the seemingly hopeless plight of our schools. Maybe the shock will motivate you to action. I want people to see the school system from a perspective that they never get to see – from the inside. Because no matter how much you are involved in your child's education, you can never know what happens on a daily basis in their lives as they roam the halls of your local school. I know, because my wife and I were about as involved as you can get. She was on the local school board, and I was on one of the high school advisory committees. I was coaching many of the high school kids in senior league, and we went to many of the high school games. One of our kids was a three-sport athlete. We were always in and around the schools for special events, parent-teacher conferences, and open house. And yet I never realized what really went on behind those closed doors until I taught school myself. This book gives you that inside look.

I had spent most of my career as an engineer before I decided that I would like to try teaching. I took an early retirement from my company, and was content to do some part-time consulting while home-schooling my youngest teenager. I didn't jump into teaching all at once. I had already been involved as a church youth group leader for a number of years. Then I was drawn in by being asked to teach Junior Achievement at the local middle school. I was also asked to judge some of the local science fairs, and then the state science fairs. I enjoyed these experiences very much, and got to interact with some of the best and brightest that our schools had to offer. I had done a lot of coaching also, and I felt I had a good rapport with the kids. I began to believe that I could make a positive impact and give back some of the things I had learned. Ultimately, I think everyone goes into teaching for this reason. Unfortunately, no one was around to tell me that I was only seeing the best kids on their best behavior in their best environment - away from school. Actually, I promised my wife that I would mention that she did try to talk me out of it. I didn't listen, which is why I am now back in engineering, but still contemplating how our schools have lost their way. How my attitude went from wide-eyed, enthusiastic, change-the-world optimism, to an in-your-face, cold slap, stonewalled realism is a story that I hope you both learn from and enjoy.

As you read these stories, don't make the mistake of thinking that your local school is any different than the ones in which I taught. My wife and other relatives and friends who are teachers, some in other states and school districts, could tell you similar stories. The problems are endemic to all of the schools, because they originate from the same sources. For most of the past generation, every survey shows that the longer American students attend school the lower they rank compared to other industrialized nations in math, science, and reading test scores. The achievement gap is widest for high school students, placing us near the bottom compared with other countries.[1] I taught in a medium sized public school (1400 students, grades 9-12) and a private school (400 students total) in a small town. Larger schools and inner city schools, I am certain, are even worse. And even though I taught only high school, the problems I encountered had deep roots which originated back in the middle and primary schools. The solutions I propose are radical, at least to the educational establishment. They may sound more like common sense to many of you who went to school a generation ago. But if you're younger than that, and have children, I hope that after you've read this book, you will reconsider sending your child to any school, particularly a public school, until lawmakers and school officials come to their senses. In the meantime, laugh at some of my stories, lament at others, reflect on what I have to say, and then ask yourself whether you want your children or grandchildren to be abused by the educational bureaucrats any longer.

Let me also say that this book is not a research project, however, that doesn't mean that my conclusions haven't been well considered and thoroughly researched. There are plenty of serious peer-reviewed studies that will back up what I have to say if you're inclined to check my references and do the homework. My experiences were real, and the conclusions that can be drawn from them should be self-evident to all except those who would hide their head in the sand. This book is simply a story about my experiences, and my take on them. The opinions expressed about certain administrators, teachers, or students are mine alone, and not intended to demean any particular person or institution. I am not a professional educator, and for that I am grateful, because it gave me an insight that those who have spent their entire careers in education don't have. As an engineer, I've taken the approach of gathering facts, analyzing them, and deducing the most efficient solution from the evidence gathered. I would like to think that this tried and true problem solving technique is superior to the emotional, "compassionate" approach the politically correct crowd has saddled us with for many years. That's what got us into this mess in the first place. "**Set Our Children Free**" isn't just a catchphrase. It is a call to action to rescue a corrupt educational system.

DEDICATION

This book is dedicated to my wife, a teacher who tried to talk me out of teaching. She was both right and wrong. This book is also dedicated to many of the wonderful, exceptional kids who were part of my life for much too short a time – the kids of great character and morality with whom I developed a special relationship. You were the greatest part of my teaching experience and you will always be a part of me, as I hope something I've said will remain a part of you. You know who you are. I may forget many of my students but I will never forget you. You are the part of teaching that I will always miss. God Bless You.

CONTENTS

	Foreword	iii
	Dedication	vii
1	Baptism by Fire	1
2	Academic Anemia	13
3	Sexual Saturation	57
4	Behavior and Discipline	99
5	The Profession of Teaching	147
6	Public vs. Private vs. Home Schooling	178
7	The Teacher as a Coach	202
8	Political Correctness	221
9	How We Can Fix It	250
	About the Author	280
	Endnotes	281

Set Our Children Free

CHAPTER 1

BAPTISM BY FIRE

My wife told me that teaching would be hard. I've never forgiven her for making it sound that easy. It wasn't hard, it was borderline impossible. Oh, it didn't seem that way at first. I taught mostly ninth graders that first year and they came in fresh and eager to start their high school career, half excited and half scared. I had already met some of the students a few days earlier, as the school permitted parents to bring their kids around to meet their new teachers. I explained to the parents that I was an engineer who was about to teach school for the very first time. No point in trying to fake it. I still remember the very first students I met, Kelly and Marti. Both were from good families, and both turned out to be honor students and outstanding track athletes. Kelly would eventually be elected prom queen. They seemed so innocent and docile, that it lulled me into thinking that all of the kids would be like them.

The first day of school went so well. The kids were quiet, did their work, and were so well behaved. I only had two classes and a planning period that day, and I thought, "What a great job!" When my wife asked me how my day went, I was happy to tell her it went well. Although she was glad, I sensed a tinge of disappointment and maybe a little jealousy, because it had seemed much harder for her when she first started. Little did I know that the first day, and even the first week or two of school is the easiest. Also nobody warned me that ninth graders are the worst group to teach by far. They are a veritable freak of hormonal nature placed into a school environment where they are the lowest freaks on the totem pole. It is akin to placing a Tasmanian devil in a room full of Bugs Bunnies. They spend all their time spinning around, bouncing off the walls, making unintelligible sounds, and when they stop, the upper classmen are standing there saying, "Eh, What's up doc?" The first day of school I taught periods 1, 2, and 3, and with block scheduling, the second day was to bring periods 4, 5, and 6 with a new group of kids. The morning classes on the second day were pretty much problem free also, although I sensed that the kids were getting more comfortable in their new environment, losing their fear, and getting bolder. It was kind of like the rumbling of a distant thunder as the sky was darkening. I intuitively sensed something was coming that I would have no control over. After lunch of the second day, I only had one class left that afternoon – the sixth period. My instincts had told me that a thunderstorm was coming, and now I was beginning to hear thunder and see lightning. Only it wasn't an approaching thunderstorm. I was about to experience a full-fledged Category 5 hurricane! When I went home that night and my wife asked me how it went, I referred to the sixth period then, by a name I have called it ever since – the SIXTH PERIOD FROM HELL!

She actually seemed pleased that I had encountered my first bump in the road, although it was more like a mountain, not a bump. Reality had set in. Shock was to follow. I had never, in all of the years I worked with kids in other venues, encountered such a pack of incorrigible, no-conscience, low-life rebels. If General Robert E. Lee had had my sixth period at his disposal, the Confederate flag would now be flying over our nation's capital – after they killed him first. It is difficult to describe the feeling of helplessness a first-year teacher experiences in a classroom full of teenagers who have no interest in your agenda, and are doing their best to make sure you don't interrupt theirs. The other periods had their share of bad actors too, who would eventually come out of the closet during the school year, but they were in the minority. The sixth period had no minority or majority. They were all just an evil incarnate horde of demon possessed, low-life scum (the exaggeration is only in type, not degree). From the moment they walked, ran, or chased each other into the classroom, until they left, they considered the teacher and any instruction that was going on as window dressing – i.e. it was something they knew was in the background but had no relevance whatsoever to what they were into. It was their world and they were allowing me to live in it, as long as I didn't interfere. The three worst students were all girls, all named Kathy, I presume because Beelzebub and Lucifer were either already taken, or too hard for their parents to spell – if they had parents. I'm still not convinced that they didn't escape from a deep dark portal in the universe with a signpost that said, "Abandon Hope All Ye Who Enter Here," when someone accidently left the door ajar in the space-time continuum. They talked incessantly, like they were on speed, never to be interrupted by the lesson, and always resentful and resistant at my attempts to focus their attention on anything other than sex, drugs, and rock n' roll. There were many days after the sixth period was over, that I just put my head down on my desk and collapsed from a combination of discouragement, frustration, and exhaustion. One time I actually passed out or fell asleep, I don't know which, until another student came into the classroom and shook me, worried that I had had a heart attack or had been attacked. It may have been either, now that I think about it. I have never had a class like that since, although what I learned that first year helped me learn how to deal with such students. Some teachers would tell me the classic advice that it's best not to smile in front of your students until at least Christmas. The reasoning is that when the kids lose their fear of the teacher, they will take advantage of you. This would not have worked for the sixth period since they never paid attention to me anyway. And it wouldn't have mattered how mean I looked, they had no fear of consequences, since even Satan himself was unsuccessful in keeping them at home.

How did I get into this mess? I was a semi-retired engineer, who was doing some part time consulting and part time home schooling and otherwise enjoying life. As I described in the Foreword, there were a series of events over a number of years that enticed me into teaching. I didn't want to apply for a position in the county where I lived because my wife had been an elected school board member there for 12 years. She was now employed as a teacher in a different county north of us, so I went to the closest high school south, and applied for a teaching position at Hancock High School. The assistant principal, Mr. Wagner, who took my application, interviewed me, and liked me well enough, but the only opening he had was a remedial class, and he wisely decided that it wasn't the right fit for me. Hancock was the only school to which I applied, and I really wasn't interested in teaching anything but high school, so I continued with my consulting business. I didn't give it another thought until the following year when Mr. Wagner, who had kept my application, called me out of the clear blue, and asked me whether I'd like to teach Algebra. His call surprised me, but I did take the position. Since I still had some previous consulting contracts to fulfill, I let him know that I couldn't just drop those commitments and that occasionally I would have to leave for short periods of time to travel out of town, which was OK with him. I didn't want to fully commit to a teaching career and abandon some high paying work, until I knew that teaching was what I wanted to do. I started the following week with a series of orientations for new teachers. It was exciting, like it always is when you are learning something new, and I couldn't wait to get started.

One of the reasons I was able to land a teaching position is that there was and is a critical shortage of teachers in math and science. Our state's Department of Education encourages those from other professions to go into teaching by publishing which courses can be taught by certain professions with a minimum of additional credits. Engineers qualify to teach math and science, and if I passed the state subject-area exam for math and physics, which I did, I could teach both. I did have to suffer through a few additional credit hours of teaching courses, giving me a first-hand look at the liberal education dogma required to be taught to all those who earn a B.A. in Education.

Set Our Children Free

One of the important things I learned in the orientation sessions, as well as my first year teaching, is that giving instruction is the easy part. The hard part is class control, which will either make or break a teacher. I will say that the head principal, Dr. Sanders was very tolerant of my mistakes the first year. I made my share, and I would not have blamed her if she had not asked me to return the following year. Some of the other teachers disliked her, but she gave me a lot of latitude, which I appreciated. By my third year of teaching I had become a veteran and had mastered many of the techniques I needed to be an effective instructor and survive. But surviving that first year proved to be too much. I had become disillusioned by the discipline problems, but that alone was not the sole reason. My biggest disappointment was in the good students that I taught, some of them honors students. Their grades fell off at the end of the year, and I sensed that they had given up. This was a trend I didn't understand, but would notice each subsequent year that I taught. My teaching wasn't any worse at the end of the year than at the beginning, but I've never been able to explain the performance drop. This was a great source of discouragement to me, so much so, that I actually resigned at the end of the year, because I felt that I had let the students down. Dr. Sanders tried to talk me out of leaving. She understood why I was discouraged, and had seen it before in first year teachers. She believed I had the potential to be a great teacher, and gave me other reasons to stay, but I had made up my mind. I have heard since then that 90% of teachers who quit, do so after the first year of teaching.

Set Our Children Free

When I resigned at the end of my first year, I had no intention of returning. I felt that I had given it my best shot, but had failed. I admit that even though I knew the subject matter, and was prepared for my lessons, and did my best to convey the knowledge to my students, I did not feel that I had accomplished anything. I even related to the students well, as long as they were well behaved and respectful. It was the disrespectful kids, the rebellious kids, the uninterested kids that I couldn't figure out, because I had never encountered any other kids like them. I didn't handle them well, and in fact, there is no good way to handle them under today's rules. Of course, these students were virtually nonexistent a generation ago, at least in small town schools. They would have been severely disciplined or even expelled back then.

I enjoyed the summer after my first year of teaching, free of the care of worrying about what would transpire the following day at school. I went back to consulting and was looking forward to some jobs I had lined up. However, I have always believed that everything happens for a reason, and the weeks following that summer reaffirmed that belief. They hired another teacher to replace me, and as August passed and September arrived, I had this gnawing feeling of unfinished business. I was never one to back off from a challenge and was never a quitter. Three weeks into the new school year, I had to go back to the high school for an unrelated reason, and talk with one of the counselors, Ms.Jones. While I was talking with her, she just happened to mention that the teacher they hired to replace me had decided to retire without any warning that same day. Hmm!

I thought this both strange and sad. It was unusual for a teacher to do something like that in the middle of a school year, and I felt it was unfair to the students. Then, a few days later I had to stop at Wal-Mart, and I purposely chose the store that was farthest from the school so I wouldn't run into any of my former students. I was there maybe ten minutes, and was checking out, when I ran into Ms. Williams, one of the teachers in our math department. The probability that I would bump into any teacher at that particular store, let alone one from our department was remote indeed. Two major coincidences in one week? She reiterated what Ms. Jones had told me, and remarked that it was a shame that a teacher resigned three weeks into a new school year, leaving students in the lurch. Because of the shortage of math teachers, a substitute, who would not necessarily even be a math teacher, would have to fill in the remainder of the year. I knew that by the time the school year had started, all of the available math teachers had already been taken, and it wasn't unusual to have openings posted for math and science teachers throughout the year. I actually felt sorry for the kids, knowing they wouldn't learn much from a substitute, and didn't think it was right for a teacher to resign in the middle of the year like that. Believe me, if I didn't have a sense of responsibility to finish what I started, I would have resigned in the middle of the first year myself, and maybe midway into my second year, as a plum consulting job became available. I told Ms. Williams to tell Mr. Wagner, that if he needed me to fill in the rest of the year, that I would do it, but I made no promises beyond that. The consulting work that I had pending had dried up anyway, and I wasn't that anxious to do more travelling away from my family to get work. Still, I was surprised a couple of days later when Mr. Wagner called, but I agreed to complete that year only. Ironically, only a month into my second year of teaching I received a call from a former colleague of mine needing an engineer in my specialty for some very high profile work with national implications. I knew that it would take a couple of months to sort out the problem he described, however I could not, in good conscience, abandon the kids again for the second time in just one semester. Even though I had previously agreed with Mr. Wagner that I would occasionally need to take some time off, I felt that two months away from school would be unfair to the kids. I told my colleague that I would travel to his facility and help him out for a week but I couldn't promise any more than that. I earned the equivalent of two months' teaching salary in that one week, but I left the equivalent of two years' teaching salary on the table to return to my school. Even though I occasionally took a few days off every year to do similar work, I had to turn down the more lucrative jobs because they required me to miss too much school. I didn't starve on a teacher's salary the next several years because my wife worked too, but because I couldn't save that much, those years were financially lost in terms of preparing for full time retirement. Like any teacher, I hoped that in exchange for the monetary penalty, I made a positive impact on some young lives.

I fully intended to finish that year and go home. A surprise awaited me however, as Dr. Sanders called me into her office, and had a meeting with me that was to extend my brief teaching career beyond one year. She said she wanted me to teach two science classes, Principles of Technology and Honors Physics, which were junior and senior level courses. I had taught freshman math my first year, and she said she hated to take me out of math, because she felt that I was going to be a great math teacher. But she said she needed me to teach those two science classes because Mr. Hicks was struggling with them and, as an engineer, I could add some real world applications to the class. Mr. Hicks possessed a dual certification like me, in both math and science, so we were in effect, interchangeable. Mr. Hicks wasn't happy to give up these classes, but I was to teach science most of the remainder of my short teaching career, as well as a math class here and there.

My very first day back on the job was so wonderful, I felt sure I had made the right decision. I was now dealing with more mature students, who seemed genuinely happy to see me back, although I had never taught any of them before. The school also had open house the evening of my first day back, which is where the parents come to meet the teachers and find out about the classes their kids take. I discovered something different right away. Unlike with the lower level kids, the parents of honors students generally show up for open house, because as a rule, they are much more involved with their kids' education. I'll discuss the reasons for this later, but the parents were wonderful, as I explained that it was my first day, but that I knew the subjects well and assured them that their kids would get the benefit of my experience. I was at peace with the new classes, and for the most part things went well that second year. I refined the way I went about my job, and by the time I began my third year, I felt confident in my methods. Oh, there were problems to be sure, but I was learning more and more about how to deal with them. And I envisioned myself teaching until my second retirement. However, Dr. Sanders and Ms. Parsons, another principal I liked, would both resign after the following year, and that was to be the beginning of the end. The two remaining principals, Mr. Wagner and Mr. Vick, both seasoned professionals, were never given a fair shot at the head principal's job, and left within a year or two themselves. They simply were not the politically correct choices, even though I felt either would have made a good head principal.

We had a new Superintendent of Schools, who appointed replacements for Dr. Sanders and Ms. Parsons. The new head principal appointee, Ms. Johnson, was a social worker by trade, not a teacher. If anyone thinks that it's a good idea to hand over the keys of a high school to a social worker with no teaching experience, I'd like to talk with them. In my opinion, we needed strong leadership at the school, not a social engineer or a therapist. Her chief henchman, Mr. Green, had been a teacher, yet he also didn't seem to relate to what was happening with real kids in a real classroom. In my opinion, all three of the new administrators - the Superintendent and the two new principals - were politically correct appointees who were given their jobs when the school board was having an extremely bad day. I'm trying to be diplomatic here, so read between the lines. I'm simply convinced that they were not the kind of leadership we needed at that school, and time would prove me correct. From the beginning, we didn't see eye to eye on several things, like academics, dress code, school discipline, and holding students accountable – nothing important! They didn't like my educational philosophy or my independence and I knew it. They didn't fire me, but they didn't have to. I left and didn't look back. Their policies took their toll on the school, in my opinion, and within a couple of years they were gone (not fired mind you, just moved to elementary schools where their coddling of students was more appropriate). After all, when was the last time you heard of a government employee fired for performance reasons? They just get recycled. It must be a "green" thing. By then we had sunk to a "D" school, according to the test scores, which is the way Florida rates schools. I'm not sure if they were replaced because of low student test scores, out-of-control student discipline problems, or both. Some of the teachers I know who still work there speculate that it was both reasons, and they report that student discipline and test scores have improved considerably since. After leaving Hancock, I taught one final year at Hillside Academy, a private school, and then I left the teaching profession for good and returned to engineering. Having worked in both government and private industry, I can tell you that the biggest difference between the two is in the area of accountability. Accountability for producing results is absolutely required in the private sector, since a company cannot survive without making a profit. On the other hand, the absence of accountability in the public sector not only produces waste and incompetence - it reproduces.

CHAPTER 2

ACADEMIC ANEMIA

One of our teachers, Mr. Roland, was flunking three-quarters of the class midway through the semester because the kids were refusing to do any work. There were about 30 students in this class, and they weren't particularly worried because they were used to sticking together and forcing the teacher and/or principal to cave in and "dumb down" the grading curve. Fortunately, Mr. Roland was one of those teachers who, as a matter of principle, insisted that his students earn their grades, since he knew that was the only way any learning would take place. Not knowing this, one of his kids taunted him by saying, "you can't flunk us all or they will get rid of you." Mr. Roland replied that he would take his chances and asked the student if he would be willing to do the same. This moment was an epiphany for the kids in that class. They weren't used to being talked to like this. The next few days approximately half of them dropped the course, which left him with plenty of space in a crowded classroom. A half-dozen others who didn't drop the class, showed up the next week with eye glasses, which Mr. Roland hadn't seen all semester, and began taking notes. Amazing how things change when students know that the teacher is serious about getting them an education. I mention this story because it illustrates two things that I observed repeatedly during my time as a teacher: (1) The students' attitudes in this class were typical, and (2) Mr. Roland's actions were not. In fact, Mr. Roland's actions were the rare,

rare exception. The student who made the remark knew this also. The reason is because, even at the high school level, teachers are held accountable for the grades they give. If grades are too low, the teacher is blamed. Most career teachers either feel helpless against the pressure that parents and administrators bring, or they actually sympathize with the students, and believe that it is the teacher's responsibility to cut the kids some slack. They will find a way, through means other than testing, to assess students so that the grade averages are acceptable. Those teachers in the tiny minority, i.e. those who hold the student responsible for their grade, fight an uphill battle because most of the other teachers have trained students to expect something for nothing. Some commentators have called this grade inflation or grade "creep" but in truth, this subject encompasses much more than the grading system. Grade creep, is in fact caused by the pressure of <u>low expectations</u>. President George W. Bush used to call this "soft bigotry,"[2] and it is exactly that. I might add that the reason Mr. Roland could afford to hold students accountable is because, like me, he didn't need the job, as he was already retired from a previous career outside teaching.

I had just finished a review of the key principles in the day's lesson, and had given the students their in-class assignment from their workbook. In-class work was essential. I learned from cold hard experience that if students didn't learn in class, very few would ever crack a book outside the classroom. Oh, I could assign homework, but it either wouldn't be done or it would be copied. The average student felt no obligation whatsoever to even think about schoolwork once they left school. So in this particular class, I had covered the material and had given a workbook assignment that would reinforce what I had taught, and which allowed me to mentor the kids while they did it as well. Unfortunately, this meant that they had to have actually listened to what I said, and then read the assignment in the workbook as well. This sounded like too much work for some of them, so Stacey, a pretty blonde junior in the front row tried to make the case why she should not have to do the assignment. When that didn't work, she decided to reach for the old stand-by excuse: "You don't teach us." I had heard this accusation before from other students. I knew that it had nothing to do with my teaching ability. The hidden meaning embodied in those four words could be characterized as a microcosm of an attitude I found prevalent throughout the school system. Interpretation: "You're not entertaining enough, and this material isn't interesting enough to hold my attention long enough to trick me into accidently learning something." Many high school students refuse to take any responsibility for learning, and this attitude is reinforced by the people who run our schools. She knew, after many years of being spoiled like this, that she was not required to give any effort whatsoever in the learning process. She knew it was my responsibility, and mine alone to cram the knowledge in her brain with or without her cooperation, while she did her nails, talked with her friends, checked her cell phone messages, and otherwise maintained her image for her "peeps." Without a doubt, this mindset was the most intractable and frustrating obstacle I had to overcome. I encountered it over and over again. Any attempt to hold the student accountable for their actions, both academic and personal, was met with fortress-like resistance.

Let me illustrate how my own perception on student accountability changed once I became a teacher. Many years before I ever taught, I was the parent of two high school students and got into a conversation with Mr. Sickler, one of the teachers at their school. The conversation was about a math teacher at that same school who was flunking 75% of the class early in the semester. I remarked that if 75% of the class was flunking, it must be the teacher's fault, not the kids'. Mr. Sickler, who would go on to become the principal at that same school, completely disagreed. I still didn't believe him, and if you're a parent, you might feel the same way, but trust me, I know <u>now</u> that Mr. Sickler was right. That math teacher was probably stubborn enough to demand the high standards that the students were not willing to meet. Believe me, I know of no teacher who wants to flunk even one student, let alone three-quarters of the class, and we need to give them the benefit of the doubt. Yes, I know there are occasional wacko teachers here and there who are poor instructors and have unrealistic demands. But I also know that school administrators usually have these kinds of teachers weeded out by the end of their first year.

One educational precept that the intelligentsia loved to recite to us teachers was, "Kids will never remember <u>what</u> you taught them, but how you made them <u>feel</u> while you were teaching it." Digest that for a moment and you will find the seeds of corruption that have infected our educational system. I believe that learning should never be about feelings or raising self-esteem. Those are problems best left to parents. Learning is about remembering facts. Facts that will become useful someday when you are forced to do something crazy, like earn a living. Unfortunately, I found, that to school administrators, the above quote is the holy grail of educational wisdom. It is all about the student and their *feelings*. If the student isn't happy, they can't learn and it is your fault as the teacher for making them feel that way.

Although this feel-good pseudo-learn by liberal educators had been afoot for many generations, it really took traction about four decades ago, and really gained momentum in the seventies and eighties. This movement had its roots in de-emphasizing cognitive academic courses like reading and math, and focusing on "affective" programs collectively known as Outcome Based Education (OBE). A common theme among these programs was the elimination of conventional letter grades (replaced by pre-determined outcomes), less-objective alternative methods of grading, group work rather than individual evaluation, a whole language approach to reading instead of phonics, a utilization of teachers as facilitators instead of instructors, and school-based health services – with or without parental consent. Many of the "outcomes" these programs were designed to achieve, presumably to improve students' behavior and classroom performance, instead, fostered many unintended consequences. The techniques utilized meditation, guided imagery, breathing exercises, and "active learning" problem-solving. Ironically, and maybe not so coincidently, these programs were becoming more popular among educators during the same time period when our students' academic performance was falling.[3] Well-meaning as these programs might have been (I'm giving them the benefit of the doubt), they had far-reaching unintended consequences, and were a violation of parental rights. At best these programs are a waste of time because self-esteem is the result of achievement, not a precursor to it. At worst, these programs infringe upon the values that parents alone have the right to instill, and cause students to reject any authority but themselves. In fact, Dr. W. R. Coulson, one of the founders of the OBE movement, now admits that these programs don't work as intended,[4] that there is in fact evidence that they may do a great deal of harm, and that he owes the nation's parents an apology.[5]

And you may also have heard about the need to teach kids "critical thinking skills." As for critical thinking, let me say that it is best left as a self-learned skill by the individual AFTER they have absorbed a minimum body of knowledge. There is a place in every school subject for rote memorization of facts and terminology. I was never taught critical thinking skills in school, and I suspect most people learned these skills as I did – by spending time analyzing the facts I already knew and comparing them to new information. Even in elementary school I learned different ways of doing math problems on my own by simply observing patterns from the problems the teacher assigned me. There must be a foundation of facts beyond dispute that will become the basis of all further learning. Once a secure foundation has been established, the mind will naturally branch out, exploring and testing new data, ideas, and concepts. This is a natural process, and valuable school time doesn't need to be wasted teaching it, at least at the high school level. I will concede that it has a place at the college level in non-abstract curricula such as law or engineering, where objective analysis of cases or problems is paramount. Outside these areas, there is no way to teach critical thinking skills, in my opinion, without introducing personal prejudice into the process. Therefore it doesn't belong in primary and secondary schools.

The feeling was almost universal among us high school teachers that something was desperately lacking in the education of the students we were getting from the primary and middle schools. Maybe it was the OBE programs, or just the attitude that engendered them in the first place, but we knew that we were getting high school freshmen who had, on the average, a 5th grade education level in reading, writing, and math. There was a ray of hope, however, at least in my mind if not most of the other teachers at my school. Jeb Bush, who was then governor of Florida, instituted his A+ education plan, which included a Comprehensive Assessment Test (FCAT) for grades 3, 5, 8, and 10, with the intent of bringing real teaching and accountability back to the schools. Failing the test could mean a student failed the grade, and the schools were held accountable for the results. Schools were graded on their students' performance on this test. Repeatedly low grades meant a school could lose some funding, and students from poorly performing schools could even transfer to another school if low grades persisted. A change in school administration was also likely. Naturally, this created a great deal of controversy on both the state and national level, but one of the great untold stories is that the plan worked. Student performance has improved significantly since the plan was implemented,[6] because teachers had to get back to teaching facts again. For the very first time, Florida students were scoring on a par with other students nationwide, and the percentage of students who scored above the national median in reading and math improved dramatically.[7]

Yet most teachers I know detested this system. They liked to freelance with students, which was a lot more fun, and a lot less painful than requiring them to actually learn something. The students didn't care one way or the other until test time came. Then the crying would start. Every year parents and teachers would complain about how unfair it was to hold their child back because they couldn't pass the FCAT test. The newspapers would be filled with articles about how much pressure the FCAT puts on students. Principals, teachers, and school board members would be quoted disparaging the test. Much of this was misplaced fear, but much of it was also calculated politics, hoping parents would pressure lawmakers to change the law. Let's get real! I have administered the FCAT to many students, and the bottom line is that most students who fail this test do so because they have either neglected their education for years, or because they are special education students. The test is not hard, and most students who have taken it feel the same way. The evidence is the pass/fail rate, and the vast majority of students ultimately pass it. And yet even the ones who pass are by no means educated to any meaningful level, since the test is a <u>minimum</u> standard, not evidence that any educational achievement has occurred. And yet, to this day, the demagoguery by the politicians, teachers, and news media continues during FCAT test week.

The problem is not that we require too much of our students, but that we require too little. School administrators constantly preach that teachers should hold students to high standards, but they never mean it. When push comes to shove, they will always cave in to parents who give in to their kids' whining. While many parents support teachers without question (this is also a mistake), there are just enough parents out there who indulge their children to the point of making a teacher's life miserable. This is not a problem with the top tier students – the honors kids who comprise the top 5% or so of every class. They are self-motivated and will do anything asked of them. And it's normally not a problem with the bottom 50% of students either, since they are in school only for the social interaction with their peers, and their parents have become accustomed to low expectations for them. But there's a large group of students between those two extremes, who are just smart enough to be friends with the top tier students, but don't want to put in the work required to get there. These honor wannabes want to be in the same classes, and thus the same social circle, with their elite friends but don't have the same motivation to learn. The problem is that in order to stay in the same classes they must get the same grades. It is also important to them to maintain the "honor student" image to their parents, who constantly push them to perform so that they can get into the best colleges. The only way for these 2nd tier kids to achieve on a level with their friends, and stay there is to: 1) cheat, or 2) whine like a stuck pig to their parents when #1 doesn't work.

If you think I'm exaggerating, I should tell you that I can think of more examples of this kind of behavior by students than any other point I'm trying to make in this book. I taught Honors Physics and other classes where I had top students mixed with the 2nd tier kids I'm talking about. I was warned by some of my colleagues that previous teachers who tried to make the class challenging were met with great resistance. They were right. Two thirds of the students in my honors classes did not belong there. These were dangerous kids in the sense that they would throw temper tantrums like spoiled children until they got their way.

Jodi was one such student. According to her and her mom, her lifetime dream was to attend a particular top university in our state, and I was standing in the way of that dream. The reason is because she pretty much had to be a straight A student, and an A in an honors class raised your grade point average (GPA) more than a regular class. She was one of a half dozen girls in that class who had no interest in the subject matter, but probably took the class because their friends did, and this was where they were hanging out that semester. Despite the fact that she expended no effort, she still expected to get an A. Her term paper was so bad that I doubt she spent more than an hour or two of her precious social time writing it. She appealed to the principal because I had the audacity to give her a C on it, prompting a parent conference with the Assistant Superintendent of the Schools who was also present. The reason he was there was because her mother had made veiled threats about filing a lawsuit against me, the school, or both. In addition, she did some research and dug up every bit of dirt (in her mind) she could find on me, and questioned my competence. Although the paper really deserved a C- at best, the principal gave in to the threats and changed the paper to a B, only because I had given the students no written objective rubric or criteria for grading the papers ahead of time. This was a mistake I would not repeat, as it was the first year I had taught the class. The semester exam was another matter, and when Jodi and some of her friends scored so low that they ended up with a semester grade of B, the battle was now rekindled with even more troops on her side determined to bring me down. They weren't about to learn the subject matter, because they never had to, having gotten their way with teachers for many years. They would sit in the back of the class and chatter the entire time, and when my warnings weren't sufficient, I moved them to separate parts of the room. This ignited a holy war in itself and I had to get the principal to come to my room and enforce my urban renewal project. As if that wasn't enough, I suspected that these girls were notorious cheaters, but it was difficult for me to tell how they were doing it even when I sat behind them during the tests so they couldn't see me without turning around. When a number of tests from students who sit beside each other come back virtually identical, even down to the wrong answers, it's not hard to figure out what's going on. I finally decided to make up three different tests, and hand them out randomly so most students wouldn't have the questions in the same order as the student sitting next to them. This was more work, but it was a practice I would continue in subsequent years, because it was one of the best tools I had to prevent cheating. The first time I tried this, I did not tell the students before the test what I had done. To my surprise, it took them about 30 seconds to figure it out even though I was watching them like a hawk (I said they were lazy and devious, not stupid). You should have heard the howls of protest. I politely informed them that if they had kept their eyes on their own paper, they wouldn't have even noticed the tests were different, and they had no reason to gripe if they were being honest. Well, Jodi did get a B for the semester, and despite her fears that I had ruined her dream of attending her university of choice, she did get accepted, and I am told, had a miserable first year. One of the girls in this clique even went so far as to try to get me in trouble for sexual harassment. When she came up to my desk demanding that she see some record of hers, I reached down to open my file drawer. Since her leg was in the way, I said "excuse me" so I could pull out the drawer. She stubbornly didn't budge, so when I opened the drawer just far enough to gently nudge her leg with the drawer, she shouted, "Don't touch me!" Of course, I didn't touch her, the file drawer had, but the students didn't know that since the front of my desk was hidden from their view. My only defense was to state, loud enough for the class to hear, that I hadn't touched her, and hope that was the end of the matter. I heard that she ended up being a lawyer. Fitting. I don't know what happened to the rest of that gang, but unless they've matured, I know that their own selfish behavior will eventually catch up with them.

I can't say I wasn't warned, but it was only my second year of teaching and I still had a lot to learn. One of the bright, but quiet young ladies who sat in the front row of the class passed a note to me after the first couple of weeks, giving me a warning about this "gang of six." She was sick of their behavior and urged me to not give in to them. This became a recurring theme in my discussions with the top students over the years. They are rarely outspoken due to peer pressure, and they are definitely outnumbered. But they are there to learn, and they are just as fed up with the behavior of their peers as the teachers are. School administrators and school boards and even state legislators need to listen to these achievers because their education is being sacrificed in order to deal with their lazy and disorderly classmates. The result is a predictable mediocrity where the intelligent students can get top grades while learning a little, the 2nd tier students can get good grades while learning nothing, and the rest of the students can get by without learning at all.

I went to a "Teacher of the Year" banquet recently for the school district in which I live, although I had never taught there. There were 23 schools in this district, so it took a long time to introduce all of the 23 teacher-of-the-year candidates from each school. During the introduction of each candidate, a lot of nice things were said about each of them, most of which were testimonials from their students. Let me just say as an aside, that some teachers actually campaign for this honor, asking kids to fill out the forms and vote for them. Whenever I was nominated, I always refused to be considered, because I knew the award meant nothing, but I'll leave it at that. Anyway, at this particular banquet, the student testimonials were meant to be complimentary, but I'll bet that none of the educators in that room realized what an indictment it was of their school system. The most common theme heard was something like, "She makes us feel so special," or "He makes class time fun," or "We can tell that she loves each one of us." Not once during the painfully long two hours did I hear what I would have expected to hear about a teacher-of-the-year – maybe something about **student achievement!** Things like, "His students scored way higher than the district average on standardized tests," or "Her students' grade averages improved dramatically over the previous year." Not once did I hear anything remotely like that. Are you surprised that teachers consider kid's feelings more important than what they learn? Were you naïve enough, like I was before I taught, to really think that imparting knowledge was the foremost goal of our school system? All the evidence says it isn't. Given that, it follows that the primary goal of our educational system is not to impart factual information, but to give kids a big educational hug so that we can all just live happily ever after in the educators' view of a utopian society. This is why an increasingly large part of each student's day is being consumed by politically correct indoctrination, and not true academics. If schools were indeed devoted to rigorous academic training preparing students for careers, instead of providing a fantasy mini-society where they can engage in all sorts of juvenile behavior without consequences, there wouldn't be time for kids to be brainwashed with feel-good programs, multiculturalism, sexual orientation training, or pseudoscience classes. Schools need to be devoted to academics, period. Employers will tell you that most of the current crop of high school graduates do not even have the minimum skills necessary to enter the work force, which is

Set Our Children Free

why many employers have taken it upon themselves to train them on their own.[8] As further proof of the dwindling role of academics, just observe how often your child's school has "field trips" loosely disguised as learning opportunities. Yes, I realize that they can learn about marine life at Sea World, but ask any teacher or student making that field trip if that's the portion of the trip that they were honestly anticipating. An honest teacher or student will tell you that they're happy that it's a day off school.

If you still don't believe that schools have abandoned academic training, let me ask you another question. Why are the vast majority of educators <u>against</u> using testing as an assessment method of student performance? You heard me right! As a whole, the educational establishment doesn't believe in testing, partly because of the influence and attitudes of the Outcome Based Education movement. If you don't believe that testing is so disfavored, then why do most schools that I know of, as a policy guideline, recommend that testing comprise a <u>maximum</u> of 30% of a student's grade average. I am not making this up - you read that correctly - 30%, and our school was no exception. If you are as shocked as I was when I first heard that statistic, then you may be asking yourself how can we know that a student is learning what is being taught? Refer to the paragraph above. Learning the subject matter is not the purpose of the school system. The purpose is: "No Child Left Behind," i.e. getting every student across the finish line, whether they have learned anything or not. After all, if one fails, we all have failed haven't we? Forget the fact that in order to get some students to pass, the education of more advanced students must suffer. It will harm some kids' self-esteem if their peers achieve and they don't. Remember, all outcomes must be equal. This is why some schools have dropped grading altogether, and some high schools have even stopped naming a valedictorian. They simply can't condone distinguishing one student above another. That wouldn't be fair.

Furthermore, we are told that some students are not good test takers, because they get stressed out, and it's not fair to them either. I cannot tell you how many times I heard a student tell me that they knew the material, but they just weren't good test takers, or because they just couldn't get the stuff to stay in their head. I will probably go to my grave believing these excuses are hogwash, not a sign of any disability as some teachers believe. Most students who use these excuses have no interest in the subject matter or are lazy. Other than a few special education individuals, who had to be tested verbally, I have never seen a student who knew the material thoroughly, who couldn't answer a few written questions correctly. Kids remember what they want to remember, and they could get straight A's if you gave them a quiz on the top 50 current recording artists, for instance. Cassie is one girl who gave me similar excuses, and then waited until the day before the final exam to ask me what her grade was. When she found out she was failing, she asked me with a straight face if she could do a book report to raise her grade. One problem. It was a math class! But she genuinely felt that she should be allowed to do (i.e. copy) some kind of book report for a grade, when she did not have the first clue how to solve a math problem because she had done nothing all year. If this kind of low expectation was an exception, it would not be a problem. Unfortunately, it was the rule! This was a repeated theme – students' expectations of passing a course while knowing virtually no subject matter. How did they come to expect such treatment from teachers? It's because they had gotten away with it so many times before. I don't have numbers for how many teachers allowed this behavior, but it must have been a significant number since this expectation was nearly universal among the average student.

So how do we assess a student's performance if we can only count tests as 30% of their grade? We are told to grade students for classroom "group" work, which most kids will copy from the few who will bother to do it. We are told to grade homework, which most kids will copy from the few who will bother to do it. We are told to grade projects and book reports, which most kids will copy from the few who will bother to do it. Is it any wonder that a good majority of high school students graduate with the equivalent of a 5th grade skill level in math, reading, and writing? I resisted these guidelines like the plague. I was foolish enough to believe that learning was about accumulating information, and the only way to see if students had learned the essential information that comprised the subject matter was to test them. After all, they are going to be tested frequently before and during any professional career that they pursue.

Despite the 30% (or lower) guideline, most school administrators allowed teachers to design their own grading system. You were free to do this, of course, only until parents and students started griping. Since I taught only science and math, I felt it was important to design a grading system that would reflect the student's knowledge of the material. After all, math and science are non-abstract courses that focus on both process, problem solving, and factual information. There is very little opportunity to ask subjective essay questions. You could grade homework and lab work, but my experience showed that there was no way to give an honest grade for either, since the kids would just copy off one another. If you didn't grade homework or lab work, however, they wouldn't do it, so it was a catch-22. So in science classes I was resigned to count tests 75%, and term papers, lab work, and homework combined, another 25% of the grade – just enough for them to do it, but not enough to raise a series of failing test scores to a passing grade. In math, I always counted tests at least 90% of the grade. A math grade has to mean something, and what it should mean is that you know how to solve the problems. Another foolish notion – you have to earn it!

As for the tests themselves, I never purposely made them hard. I simply tested what I had repeatedly assigned for homework and had gone over in class several times. To my surprise, the majority of my students actually wanted me to give them a copy of the test one day ahead of time so they could study it and memorize the answers! When I countered that my eight year old granddaughter could score a 100% by doing that while knowing absolutely nothing about the subject, they replied that other teachers allowed them to do it all the time! (I did in fact let my granddaughter try out one of my high school science tests by giving it to her ahead of time, and then testing her 30 minutes later. She got a B on the test, while literally knowing nothing about the subject matter, proving my point). Every year I taught, I raised my standards, and I found that student performance went up right along with it. That told me that every student makes up their mind from the first day of school what grade they will get in a particular class, and they will only do the minimum amount of work to get it. For most students, this is a B or a C, and many of them will settle for just passing if they decide it's too much work to get anything more. Honors students won't settle for anything less than an A, and will do whatever is required to get it. In this regard, it is no different now than when I was in school a generation earlier. I knew the grade I wanted in a particular class, and I did only the minimum needed to get that grade, which was usually nothing. If my teachers had made it harder for me to get the grade I expected, I would have worked harder, but I would also have learned more. This hurt me later on in college. Although I was an excellent math student, I had a difficult time with some of my engineering courses. One of the reasons was because of my weak background in Physics – a fundamental course for all engineering disciplines. Both my high school and college Physics teachers never challenged me to learn the subject matter as thoroughly as I should have. Therefore I didn't have this fundamental knowledge at my grasp while studying more in-depth technical courses, and I had to re-learn Physics on the fly.

No matter what kind of test I gave my own students, they always expected a little extra help during the test. This was after I made it clear that once the test was passed out, I could only answer a query from them if they did not understand a question, but I couldn't help them answer the question itself. Then, after the tests were handed out they would ask things like, "Can you help me with this problem?", or "Is this the right answer?', or "I don't understand, what do I do now?" I tried to tell them that tutoring was available after school, but I couldn't tutor them during the test. Many of them felt this was so unfair, and again, the only reason they had this expectation is because they had been spoiled by other teachers doing this for them. I knew one teacher who would give his students 10 minutes of study time in class before a test. He said 50% of the students would just talk or waste the time rather than study. When he asked how many had actually studied, 3 or 4 hands would go up.

Teachers are constantly urged by their school administrators to hold the students to high expectations. Backing up those words, however, is a different story. There was one principle that I could not compromise on – you get the grade that you earn. I designed my syllabus every year around it. It had an actual description of what an "A" was, a "B" was, etc. It described the skill level to obtain such a grade. It described the grading system and the percentage of the grade things like tests, lab work, term papers, homework, etc. comprised. The syllabus was sent home and had to be signed by the parents. When things like term papers were required, I also included a grading rubric which described how points got added or subtracted from its grade. I even included a worksheet that showed how to keep track of their grade. No student I know ever used it, since it was easier to just ask me. The point is, it didn't take a rocket scientist to figure out what was expected if you wanted to get a good grade in my class. Expectations were clear, but the student work ethic was not. Again, the above does not apply to the top students, but it does apply to many of the students, and I am not exaggerating. A teacher must dispel these student expectations early in the year, or student performance will be lacking for the entire year. As it was, the first quarter was always the most productive, followed by declining student performance throughout the year. The final quarter many of the students just quit, especially if they had a high enough average to keep them from flunking. My insistence that a student get the grade they earned was not grounded in some twisted legalistic concept of reward and punishment, nor from a lack of compassion. Like most teachers, I had to periodically ask what purpose I was serving in this profession. The answer always came back that if I wasn't imparting any knowledge, then I was wasting my time. I wasn't into sitting around in a circle with students, drinking hot chocolate, and singing Kumbaya. And in order to assure that students learned the material that I was hired by the state to teach them, I had to make sure the grading system was honest. Giving away good grades without requiring knowledge of the course content is the surest way to assure that no learning will take place. Kids are going to take the path of least resistance. They won't do any more than you require of them to get the grade they want, and many of them won't even do that much, hoping that their whining will save the day. This is why we need state standardized tests to assure accountability, or teachers will just default to mediocrity or worse.

Set Our Children Free

Most kids really feel that they should not be required to learn the course material if it is not interesting. If you don't believe me, listen to them talk about a course in which they received a bad grade. They will say things like, "It's boring," or in some other way blame the teacher's methods. They are loathe to accept responsibility for the grade. You will never hear them say, "I just didn't study." Even when they say, "I just don't get it," it implies that the teacher hasn't done their job in making them "get it." Sometimes I told the students to just make themselves do the work whether they liked the subject or not, because it was good preparation for adult life when there would be many things they would be forced to do even if they didn't want to. I'm here to tell you that that philosophy has gone the way of the dinosaur. That kind of thinking, I was flatly told by one of my principals, was totally foreign to today's students. She was right. Not only is the burden of learning now placed squarely on the teacher's shoulders, the teacher must find a way to make the class interesting or the student feels no obligation to learn. "Why can't you make it interesting?" they would say. My usual reply would be something like, "If I was an entertainer, I would be in Hollywood, not here." Recognizing that we live in a digital generation, however, I tried to find as many videos on the subject matter as possible to sprinkle in with lectures and demonstrations, etc., although this wasn't always possible, especially in math.

One of the reasons kids have to be entertained into learning is that they have never been brought up to appreciate reading. So much of the information coming into their brain is in video or digital form. I don't know where I came to appreciate my love for reading, but I'm thankful for it, and I suspect many of my generation feel the same. I would rather get my information from reading than any other venue, and to this day, I'm not happy unless I have completed at least a couple of books a month. Today's students would rather do anything than read, even though I still believe that is the best way to convey information. If I gave a reading assignment, even to honors students, I could be absolutely certain the students would ignore it. They either got the information from what I taught in class, or they didn't get it at all. I don't know why we even issue them books. So stubborn was their inhibition to reading that when we did a science lab, I couldn't even refer to the procedure given in the science workbook. I had to give them a condensed, type-written step-by-step numbered procedure or they had no chance of getting it done. Even then, each group would stand around, with the procedure right in front of them, and say "What do I do?" If I told them to read the simplified procedure I had handed out, they treated it like it contained a deadly virus. Mind you, I had already demonstrated the lab experiment for the entire class before I broke them up into groups. But each group insisted that I personally demonstrate the lab procedure to them again, rather than read the step-by-step procedure I had just given them. Even the top students had this disease. Once, in response to a question, from a future valedictorian no less, I asked if he had read the chapter I had assigned. His response was, "You're funny Mr. C!"

Lack of reading skills is a huge impediment to learning at all levels of the educational system. You may have heard about how many students graduate from high school without having learned to read, and wonder how that can happen. Trust me, none of the teachers I know are puzzled by this. They not only know how poor the reading skills are right up through high school, but they know exactly how an illiterate student can earn a diploma. Teachers are so pressured into developing alternate teaching strategies and means of assessment, that it comes as no surprise that a student can graduate without having learned how to read. Between listening to lectures, watching videos, cheating, group work, and getting another student to do papers for them, students don't need to learn how to read. The reading deficiencies start early in life because many kids didn't have parents that read to them from actual books. It is easier to let the TV babysit a pre-schooler than to take the time to develop a love of reading in the child. Parents, this is a huge mistake. One of the most beneficial things you can do for your pre-school or primary school child is to sit them on your lap and read with them. This will not only develop in them a bond with reading, but a bond with you. This is probably the most detrimental trend in the modern era, because in my opinion, learning from reading, at any age, is far superior in both quality and detail than video learning by far. Intelligent and learned people <u>read</u> – it's that simple.

Unfortunately, instead of diagnosing and correcting low reading skills in the primary schools, the problems are left to fester until high school. Our school was even foolish enough to pay $80,000 for a reading program whose vendor claimed great success in bringing reading skills up to par. After seeing the program, I was very skeptical. No one questioned the vendor's claims but me. They were shocked that I had the audacity to ask for the data. When they showed me the so-called "data," I pointed out that it involved studies of primary and middle school students, not high school. They told me they would send me the high school data, but they never did, which did not surprise me because I knew there was none. The reason I knew is because anyone could have concluded that such a juvenile program would not work on high school students. The program was so juvenile, in fact, that most of the teachers and students were embarrassed to be part of such a charade. I couldn't believe we were being asked, for instance, to read stuff only a little more advanced than "Fun with Dick and Jane," and then having the students answer questions about the material at the snap of our fingers (literally!), which was part of the program. Mind you, we were required to take a portion out of our class time in academic subjects such as math and science to do this. Forcing this idiotic program on us was one more decision that caused me to question the competence of the administrators who were running our high school. I know what you're thinking. If a kid has reached high school and still doesn't know how to read, or has very low reading skills, then why not put them into a concentrated reading class for a semester, instead of taking an elective? Why require math and science teachers to take valuable class time away from all of the other students, in order to teach the just the ones who don't know how to read? If you think this way, you obviously are possessed with a great deal of common sense, and therefore don't belong in the education business. You see, identifying and singling out the low readers could make them feel bad, and obviously that would be <u>much</u> worse than not teaching them how to read. The other kids in the class who knew how to read would just have to put their education on hold and suffer through this sham. It was a small sacrifice they would have to make for the good of all. Remember, No Child Left Behind. We also had to circumvent our wonderful state laws that prohibit a student from being taught to read more than a certain number of minutes a day unless the instructor was certified in that subject area (reading). Since we didn't have enough certified reading instructors, we all had to suffer together. Forget the fact that we let substitutes and even regular teachers teach courses for which they hold no certification, for as long as an entire year. Although most parents, including myself, oppose year-round schools, I still think that schools should require students to read a certain number of books during the summer. Those who lack reading skills should be required to take mandatory summer classes until they can read at their grade level.

As for group work (excuse me, I mean "cooperative learning"), we were told that it was important because corporations supposedly value an employee's ability to work within a group. Having come from the corporate world, I knew that this was taken completely out of context. Of course, in a corporation you have to work with other people as part of a larger group, but trust me, you are evaluated as an individual as to your worth to the company. Yes, there are times when you are rewarded as a group, but your value to the organization is always measured by your individual abilities. It's scary to think that for many of these students, college or the workplace may be the very first time in their lives that they will be personally held accountable for what

Set Our Children Free

they produce. Also hastening the move to group work the past couple of decades was another school "reform" called block scheduling. There are many variations of block scheduling, but a typical school day would look something like this: Instead of 7 or 8 classes per day, approximately 50-60 minutes each, block scheduling would have only 3 classes per day, two in the morning and one in the afternoon. Each class would be approximately 1 1/2 hours long to allow for group activities and labs, which the old schedule didn't accommodate well. Teacher planning periods are longer under block scheduling also, as well as the time spent between classes (20 minutes vs. 5). Since a typical student schedule consists of at least six classes, with block scheduling each subject can only be taught every other day, whereas in the old schedule, each subject is taught every day. Most educators love block scheduling, and therefore credit it for all sorts of benefits from better student behavior to solving world hunger. I am of the opinion that block scheduling is one of the worst reforms ever to hit the school system. A small but sensible minority of teachers feel the same way, particularly those who have taught under both systems. First of all, with block scheduling, the classes are so long that there is no way to hold the students' attention. Therefore a teacher has to resort to inventive ways to keep the students focused. We are, in fact, told to change the direction of the class three times within those 1 1/2 hours – lecture, lab, group projects, for instance. This results in much more wasted time because students can only absorb so much material in one class lesson, so the lesson has to be watered down to fit within the time period. To make matters worse, a student won't see the same teacher or subject matter for another two days. If a student has Algebra on Tuesday, for instance, they won't have that class again until Thursday. If a student is absent on Thursday, they will have gone six days, until the following Monday until they are exposed to the material again, and absenteeism is a huge problem. And because not as much material is covered, since classes are only held every other day, a student will only earn a half credit per semester, instead of a full credit under the old system. I have taught under both systems, and I can tell you of a certainty that students exposed to a small amount of material every day will retain much more than when they are exposed to a larger amount every other day. And less review is needed with the shorter class periods. I had no problem giving tests and performing labs under the 7-8 period schedule, contrary to those who say there is no time to do either. And the 20 minutes between classes is too much time for the students to get into trouble. The old schedule with 5 minutes between classes makes for a busy day with little time for nonsense. From the student's point of view, the mood is one of learning, since there is little time for anything but going to class and preparing for the next one. When the last class is over, they get on the bus and go home. Other than lunch, there is little time for anything but academics, and that is the message that the schedule should convey. The reason we have block scheduling, in my opinion, is because teachers want it, not because it's best for students. There are studies that back up what I am saying,[9] but I'll admit that no one has done a study using the same group of students, taking the same class, using the same book, and the same teacher, for the same amount of time, under both schedules, since that is impossible. But I have taught the same subject from the same book under both systems, and I was able to easily cover more than twice as much material per semester under the standard 7-8 period day schedule. I think any objective observation by a teacher using both schedules would conclude the same, looking at all of the factors which contribute to learning, (focus, repetition, self-discipline, etc.).

One reason for poor student performance is absenteeism, and making matters worse (big surprise here), poor students miss more school. This is so true it is almost a proverb. Is the former the result of the latter, or vice versa? I can tell you what I have observed. Top students never miss class, good students rarely miss, and poor students miss often. They miss because they lack motivation. They have such terrible home situations, that getting through the school day is the least of their worries. Even for those who come regularly just to get away from home, there's no guarantee they will attend class. Since all absences are reported to the parents the same day, it is difficult for a student to talk their way out of it when they get home. Because of this, I have seen students who had near perfect attendance records, but because they slept through every class, they failed most of them. On the other hand, I have never seen a top student who had more than one unexcused absence per semester. In order for a student to excel, they must attend class. Even top students, who have legitimate reasons for missing at times, suffer some penalty when they are absent, because it is impossible to completely make up all of the instruction they missed. The more marginal the student, the more they need to be in class, yet the more likely they are to miss.

Most top students never miss at all except for an excused school function. With block scheduling, anything over six unexcused absences means a mandatory failing grade for that semester. This was a district-wide rule, but as usual, exceptions were made depending upon whether the student was a top athlete or well liked or near graduation. Dropouts, expulsions, discipline problems, and graduation rates are all tracked by the state and some federal agencies, and can lower your school's grade. A low school grade brings bad consequences, as discussed earlier, leaving every incentive for school administrators to tweak the numbers.

And when kids do miss class, many of them are very selective about which days they miss. The day of a test was far more likely to be missed than any other day, and it didn't take long to figure out why. I knew a teacher who had two students who were present every day of the entire school year **except for the test days**. True story. My first year of teaching, I was naïve enough to grade the tests and give them back to the students the same day. If a student was absent, they would get a copy of the test from a student who took it the day before, so they knew what the questions were. It didn't take very long for me to figure out what was going on – after all, some of the kids would be stupid enough to bring the graded and borrowed test to class with them, so they could copy the answers while taking their own test. They were so casual about this that they didn't even try to hide it. From then on, I started making up two tests – the original and a make-up. Because some students missed even the make-up day, I finally abandoned this practice also, and just started collecting all tests and not handing them back until everyone had taken them. This was also difficult to do because most students strenuously objected to taking a test the day they got back from their vacation, ...er I mean sick day. Even though I made it a rule in my syllabus, and re-emphasized it over and over again throughout the semester, it was still the source of a great deal of contention. The students honestly felt that they should get a free ride the day they missed **plus** the day they came back. I couldn't review the test until all the students had taken it, and I couldn't do the review with the students present who hadn't taken the test yet. This meant that I had to put off the review for a week or more if the absentees didn't take the test the day they returned. This wasn't fair to the other students since many of them wanted to go over the test to see which questions they got wrong. Waiting a week or more meant we had already moved on to other material and the students no longer cared about a review. This is one of many instances that I could cite where bad behavior by some students was harmful to the education of others who really wanted to learn. See Chapter 4 for many other examples. Add to this the fact that most students refused to take make-up tests on their own time, i.e. lunch or before or after school. This meant that they not only missed the day of the test, but also another instructional period to take the make-up test, making absenteeism very costly indeed for them as well as the other kids who didn't miss school.

By my second or third year, I alternated between several test methods, always with the goal of fairly assessing what the students had learned. In many cases, I just made up two or three tests with different questions, or the same questions in a different order. Then I would give a different test to each row of students in the classroom, since it was easier to cheat side-to-side than front to back. Then I would TELL the students before the test that I had done this, and to not look on their neighbor's paper because it was unlikely to be the same test as theirs. Despite this, I always had at LEAST one student in each class turn in their test with all of the answers to their neighbor's test. I would look at their test and say, "Hey, those are all correct answers." The student would smile, and then I would say, "Unfortunately, they're the answers for your neighbor's test, not yours," as I engraved a zero onto their test. I've never figured out what went through their mind because I know they heard my warning. They must have either thought I wouldn't notice, or that they would take a chance on winning the lottery, since they didn't know the answers anyway. Speaking of cheating, you would be shocked to learn that I got a great deal of resistance to giving zeroes for cheating. I was urged by more than one principal to either: a) give the student a 59% (one point below passing) on the test, or, b) to let them re-take the test. Since I was not about to give a 59% for doing nothing, nor penalize myself by having to make up another test, I chose, c) none of the above – they got a zero.

Another testing technique I used was to make the original test a multiple choice format, and make all make-up tests a fill-in-the-blank or essay format. This seemed to miraculously cut down on the absenteeism on test day. As I said before, the students were ingenious at finding ways to cheat. I would sometimes watch suspected cheaters the entire test and still not have a clue as to how they did it.

Set Our Children Free

Most of the time, I would call the suspected cheater up to the board and ask them how they got a particular answer. This was a good technique for weeding out math cheaters, because if they couldn't show me how they arrived at the solution and do it over for me on the board, the answer was marked wrong. In math, I also insisted that everyone show their work right on the page, or the answer was wrong. I rarely got challenged on this, which told me that the student knew what I knew – they had obtained the answer dishonestly. If this sounds cruel to you, then I would have to ask you whether it isn't more cruel to let a student get away with cheating while learning nothing, or to insist that students behave honestly and not only learn the subject matter, but a lesson in morality to boot. All laws and rules have some kind of moral root, and when they are not enforced, immorality will run rampant. Our entire legal system is based upon telling the truth, which is why perjury is a serious crime – unless you happen to be the President of the United States. Similarly our entire economy which operates on oral and written business contracts, would collapse if people didn't honor their word.

The law is a teacher and a morally restraining force, and it is best that kids learn that at an early age. If there were no sanctions against dishonesty and lying, where would the limit be? You wouldn't be allowed to be fired for lying on a job application, and you could fraudulently sell any product even if you lied about it, or it was defective. Business transactions would grind to a halt because nobody could be trusted to engage in even elementary commerce due to lack of trust. And lying, immoral politicians would occupy the highest offices of our land, with the full complicity of the news media, who had forsaken their moral role as a government watchdog (Oops, we're already there!) - which proves my point, by the way. There are those in our society who want to legalize drugs, prostitution, and gay marriage because, they say, we can't stop the behavior anyway, and people should be allowed to do what they want as long as they don't hurt anyone else. For those who so advocate, I dispute the notion that legalizing such activity would harm no one but those involved. Since the law is a teacher and a morally restraining force, how many people, absent such restrictive guidance, would engage in those activities if they were legal? Experimentation always follows legalization, because legalization implies official authorization. The 20's era Prohibition is always given as an example of why such laws don't work. I would give the same example to argue that many millions more people took up drinking since prohibition ended. The result has been an untold number of lives shattered by violence, rape, murder, and alcoholism, not to mention millions killed or crippled from car accidents. The lifting of all sanctions on personal morality results in anarchy, not freedom. Benjamin Franklin himself said, "Only a virtuous people are capable of freedom. As nations become corrupt and vicious, they have more need of masters."[10] John Adams said, "...Our Constitution was made only for a religious and moral people. It is wholly inadequate for the government of any other."[11] Our Founders gave us a system of self-government. When "we the people," i.e. the "self" part of self-government, descend into immorality, we are no longer capable of governing ourselves and we will willingly submit to slavery.

Set Our Children Free

Before I went into teaching, my wife told me that a majority of kids at her school had no interest in learning, but only attended school to be with their friends. I had a hard time believing it, but after having taught school for several years, I can affirm her statement wholeheartedly. The truth is that high schools are a miniature society where everything is acted out very similar to an adult society, with the exception of the part about having to make a living. A student going to class is just like an adult going to a job every day that they hate. They go "home" after class or during recess to their significant other. They have fights, get divorced, cheat on each other, and bring the same morality to the classroom that they will to their job when they are adults. They drive cars, have sports heroes that they worship, and go to parties. They have friends with whom they engage in athletics and extracurricular activities. They drink, do drugs, and go to their "job" (class) the next morning stoned. They have drama, broken hearts, and unwanted pregnancies that threaten their job status. They fight with their boss (teacher) and get upset at their job evaluation (grade). If it gets bad enough they have to change jobs (schools). How have our schools turned into this reality TV series? Because those who run our schools not only tolerate it, they actually encourage it. They give lip service to the notion that they are running an educational institution, but their attitudes and actions speak otherwise. They love the homecomings, the proms, the football games, and the dances, but they have no stomach for the tough choices necessary to change the culture. I noticed this many years ago with college presidents in the era of student protests. The college administrators were notoriously wimpy, and always gave in to the student demands. I thought, "Who's being paid to run this institution?" Well, high school administrators, as a whole, are no different. They don't have to put up with student demonstrations, but there's no question who runs the show. "We are here for the kids," is an oft-repeated phrase at school by everyone who is not a kid. Am I the only one who thinks this is bad policy? Today's kids fully understand that this is the school's official philosophy – we're here for you. They are spoiled at home and indulged at school to a degree unknown to the previous generation. This attitude infects everything from discipline to academics.

A very frequent cause of interruption in my classes would be a drama related to one of my female students' personal lives. It would start with one or more girls in the class asking to be excused, or maybe several girls from outside my classroom interrupting me in the middle of instruction and asking if they could talk with someone in my class. When I inquired about the reason for such an interruption, they would say something like, "Stacey's crying and upset in the restroom," or "Brandi's having a hard time right now and she needs me." If I dared inquire further, I found that it always had something to do with a boyfriend. I would always say no to these requests, and the reaction from the girls ranged from disbelief to outrage that I could be so coldhearted. I would point out that this was an educational institution, not a lonely hearts club. Sometimes the girls would ignore me and leave anyway. To them, as well as the majority of the students, anything going on in their personal life was far more important than their education. It didn't matter if they missed some instruction, but the real life drama could not wait until recess. Another contributor to this insanity is Hollywood. They produce entertainment that depicts the average high school student doing exactly what I have described above. In this case, life really does imitate art. Movies and TV wouldn't be entertaining if the plot were about education instead of relationships. This is one reason, among others, that students expect to be entertained into learning.

Instead of teaching the kids that education is a privilege, and that they are there to learn, we teach them that <u>we</u> are there to please <u>them</u> and make them happy. Because of this attitude, a teacher cannot challenge students academically like they should. There are simply too few administrators who will stand behind them. It could cost both the teacher and the administrator their job because the problem goes even beyond the school level. As I said earlier, state and federal laws penalize high dropout rates, high discipline referrals, low graduation rates, and a high number of expulsions. They even penalize you if too many discipline referrals are minorities. A nationally certified Louisiana teacher recently sued her school because she was told that she could not fail her students under any circumstances, even if they deserved it.[12] The only thing newsworthy about this story to most teachers is that this particular teacher actually brought a lawsuit. Most teachers know that there is an unwritten rule that they are not permitted to fail more than a handful of students, or their competence will be questioned.

But what can a teacher do when a kid is unmotivated? You can be compassionate all day long to kids whose home life is a nightmare, but sooner or later you have to get them to learn, and you can't motivate where there is no motivation. Coaches are not expected to do this with their athletes, and yet we are expected to motivate students. Can you imagine a principal blaming a coach for a lazy, unmotivated, yet talented athlete, who was underperforming? If we don't do this with coaches, then why do we do it with teachers when their students won't put in the work? Many teachers find a way to pass students by lowering their standards. I don't have a problem lowering standards for students with actual learning disabilities, but disinterest and laziness are not disabilities. As for students with actual disabilities, I never had much problem with them. Many times they were more motivated than the lazy students, but there were a few who expected to be given their grade because they had been conditioned by the law to understand that they required special treatment. By the law, I mean the Americans with Disabilities Act (ADA), a well-meaning piece of legislation, but the poster child for unintended consequences. The unspoken rule among teachers is that you could not fail an ESE (special education) student once they were so classified under the ADA. Teachers could be, and have been personally sued under this Act for not providing all of the accommodations a student required. Once a student is classified as ESE, their effort is no longer a factor in the grade they receive, just whether the teacher has done everything possible to accommodate them. Of course some students took advantage of this, and many teachers just found it easier to pass these students than to fight the system. The

Set Our Children Free

ADA is also one reason why many such students cannot be suspended from school longer than 10 days, even when the rules violations involve drugs or violence or things that would get a normal student expelled permanently.

Don't be deceived by those who trot out the best and brightest students to prove our educational system is fine. One example that comes to mind is the show "Are you smarter than a 5th Grader?" Trust me, the students on this show are way above average. I know this because they are smarter than many of the high school students I taught. If you don't believe that educational standards have declined over the years, read the eighth grade final exam from 1895 which was published by the Salina (Kansas) Journal in 1996, and documented that same year by the Smoky Valley Genealogical Society. It can be found on the internet at the referenced website in the endnotes at the back of this book.[13] Read this test and you will conclude that 90% of <u>college</u> graduates could not pass this test today. There are those that would argue that this exam tests specific knowledge, and our schools today offer a much broader based education. I completely disagree, and I could make a very valid argument that this test measures knowledge that all high school students should know. But let's assume for a moment that there is validity in the argument that it is a specialized test. I can tell you most assuredly that I could spend an entire semester just teaching only the information needed to pass the 1895 test, and a majority of today's high school students would still fail it. Remember that the 1895 test was given to **7th graders**, who were allowed to re-take it again in 8th grade if they didn't pass it. Also, the school year was only seven months long. Why am I so sure that most high school students would fail it? Several reasons are apparent. Passing a test of this type requires reading and retention of knowledge, repeated practice of math skills, and good writing and composition skills. Most high school students today would refuse to work hard enough to read, retain knowledge, and practice the skills necessary to pass this test. In addition, their writing and composition skills are so poor, they would require several semesters of remedial work just to properly answer the many essay questions. In today's video and digital generation, kids are rarely required to read and write anymore. And if you don't really believe the standards were that high in 1895, then many of you baby boomers can simply hearken back to your own school years a generation ago. I don't think there's any question that the academic standards were much more rigorous then than now, and likely more rigorous the generation before that. Standardized test scores seem to back this up, but since there are a number of other factors that influence these scores, they alone cannot be used as a reliable indicator. One thing is certain. One or two generations ago, functionally illiterate students rarely survived 12 years of education without dropping out. These days they graduate with a diploma. All they have to do is occupy a seat, and some educator will find a plan to get them through. The top students today, however, would have done well in any generation because of their makeup and motivation. They would pass the 1895 exam because they would do whatever is required to learn the material. Unfortunately, as I said earlier, they compose only about 5% of the student body at best.

Another indicator that tells me I'm right is the quality of foreign exchange students I have taught. I've had kids from Europe, Asia, and South America who studied right along with my best students in Honors Physics. Even with the language barrier, most of them were one or two grade levels more advanced than their American counterparts. In fact, the Physics teacher who replaced me when I left told me that he had a foreign exchange student from Serbia, who showed up the last day of the first semester while his students were taking the semester exam. He gave the Serbian student the exam just for fun to occupy him while the other students were completing their exams. His students (remember these are honors kids) scored mostly in the 50-60% range on the exam, while the Serbian student, who had not been there all semester, scored 88%. When this teacher asked him how he managed to do so well, the student replied that he had been studying Physics since primary school! I heard similar stories from my own foreign exchange students.

One more example is how our grading system has been watered down. I still remember a generation ago when the following grading system was used in most schools throughout the country. Later, states were pressured to standardize and weaken the grading scale so students could compete equally for scholarships, although some states still use the 1950-1960's scale shown. Compare that with the grading system of most schools today.[14]

Set Our Children Free

1950-1960's		Florida 1987		Most States Today	
Grade	Percent	Grade	Percent	Grade	Percent
A	93-100	A	94-100	A	90-100
B	85-92	B	85-93	B	80-89
C	77-84	C	75-84	C	70-79
D	70-76	D	65-74	D	60-69
F	< 70	F	<65	F	< 60

I would like to add a final positive note about a small group of kids who don't fit into any of the preceding categories I've discussed. Every year our school gave an award to a bad student who had turned things around. I knew several of these students, although I can't take credit for their rescue. One of them was a lovely girl I'll call Linda, who I had as a student when she was a senior. She told me that while she was growing up, her father was one of the biggest drug dealers in the county. She said he was currently in jail, but that one of her earliest memories was seeing bags of dope everywhere she looked in her house. After he was arrested, she was left with an aunt who partied a lot and took Linda with her. By her freshman year, Linda was well on her way to destroying her life also, and no surprise, was a horrible student. Joe, a teacher friend of mine would observe the boys in her class make fun of her for being stupid. He called her in one day, and gave her a way to get even with the boys. He told her that if she came to his room and studied, he would guarantee that she would pass the tests. She did just that, and began to show the boys up with better grades. More importantly, it gave her the confidence to start taking school and life seriously. She did so well in her final three years of high school, that she raised her failing freshman average to a cumulative 3.0 by the time she graduated. She would go on to college and be a model student and citizen. Another success story that I was personally familiar with was John, who was one of the most devious, twisted kids I had ever met. He was in one of my freshman math classes. He was a pervert with a filthy mouth, and did everything he could to make my life miserable, all the while smiling like the cat that ate the canary. To make matters worse, he wasn't stupid, and he would devise the most ingenious schemes imaginable to cheat on my tests. In addition, he was disrespectful, and I probably wrote him up and threw him out of class more times than any other student. So when he showed up two or three years later in another one of my classes, I groaned inside, and started thinking about how to pawn him off on another teacher. To my pleasant surprise, however, he had totally changed. He now had a quiet, confident, respectful spirit that was obvious to me from the first day of class, and he was now taking his education seriously. When I had the chance to talk with him one-on-one, which I took time to do with every student early in the semester, I discovered why. He had become a Christian. The transformation from troublemaker to trouble-free was so complete, that the word "miracle" was not too strong a description. I was reminded once again that one of the greatest solutions available for reforming our schools, is the one that is purposely banned by law from being used on school grounds – the life-changing formula found in the Bible. Somewhere in the past, John apparently didn't listen very well in biology class when they were discussing the state-sponsored religion of evolution. Good thing he missed the part about there being no Creator, or he might still be the same low-life scumbag that I used to know.

CHAPTER 3

SEXUAL SATURATION

I was teaching a lesson on radiation, and explaining that everybody is radioactive. This is because we all have traces of radioactivity in our bodies from isotopes of the elements carbon and polonium, among others, that we ingest while eating. So we actually get very small amounts of radiation from everyone we are near. It's one of many sources of background radiation to which we are exposed every day. The total radiation exposure we get, and thus the danger of it doing any harm, is directly proportional to amount of time we are exposed to it, and the exposure we get from other people is very small. I went on to say that even married people get only the equivalent of several chest X-rays of radiation from each other over a period of a lifetime. This was a little too much for some of my students who were more than prone to make a sexual joke out of anything the teacher said anyway, even if the discussion had nothing to do with sex. Neil was one such student, although calling him a student was exaggerating a little. He did virtually nothing in class, and the word was that he was attending school just to play ball. He raised his hand, and I knew what was coming – a sexual innuendo to distract the class and waste time. He started his question with a knowing smile, and within two seconds the students started chuckling because they knew where it was going. He began his query with, "How much radiation exposure would I get if I was engaging in *close, very close, intimate* (his emphasis) contact with someone of the opposite sex, for a time period of say ......" At that point, I cut him off with, "one minute?   No, that won't hurt you!" Hearing that, the rest of the class exploded in laughter which didn't subside for quite a while. Neil, now a nice shade of red, was speechless for once.   Thereafter, the students' nickname for him was "Minuteman." I honestly didn't intend to embarrass him and just wanted to get back to the subject, but sometimes, you know, radiation happens! My surprise at the students' reaction was not because they laughed at a joke about sex, but that virtually an entire class of 16 year-olds understood <u>why</u> it was funny.

A teacher cannot say anything that has the slightest chance of having a sexual connotation, without getting giggles or comments from the class. I can't tell you how many times I said something I thought was totally innocent and unrelated, only to see the smiles and laughing, until I realized that, yes, I guess you could take that phrase differently if you really stretched your imagination. Their imagination however needs no stretching. Kids spend more time thinking and learning about sex than any other subject. It is every student's major in high school. You could say that they major in sex, and every other subject is a minor. And why wouldn't they? They are saturated with it. They are primed every night with sexually saturated TV shows, song lyrics, commercials, Hollywood gossip, and movies, only to come to class the next day and hear tales from their fellow classmates, which they are more than willing to repeat and discuss. All the while, the hormones are raging inside of them.

Set Our Children Free

I can't tell you how many girls came to my class with a Cosmopolitan Magazine – not Cosmo-girl, which is written for teenagers - the other, adult version! They would insist that they weren't reading it for the sexual content, but I would point out that the magazine is pretty much devoted to that subject. I would even take the magazine from them and read out loud just the titles of the articles on the cover. They would laugh and be slightly embarrassed, but then change their tune and say they know all about that stuff anyway. Their idea of a good time is to spend Spring break at Daytona Beach with a group of their friends, without parents or chaperones. I wonder how many of the guys they hook up with know that they are having sex with underage high school girls. I was also shocked at how many of the girls in my class went to the Gasparilla festival in Tampa and "flashed" for necklace beads thrown from the guys on the parade floats as they pass by. Stacey, one of the girls in my class, who told me that she enjoyed flashing at the Gasparilla festival, seemed excited when she told me that a picture of her escapades was making its way around the school. This has become a crime in some states, as many of these girls, like Stacey, are underage.

One girl named Sherry used to brag that she had been with a hundred and sixty guys. And she was just a junior at the time, if I remember correctly. A lot of the girls were ticked off at her because their boyfriends or ex-boyfriends had been with her, so I assume the lack of character concerning this girl, if not the number of sexual conquests, was correct. And it used to be only the boys who kept score. I remember wondering what kind of future she had as I watched her walk across the stage and get her diploma. I also wondered what kind of family influences, or lack thereof, could have created this social derelict. And finally I wondered how she even managed to graduate given that she couldn't have spent much time on academic pursuits. It just goes to show how easy it is to get a diploma without getting an education.

Electromagnetic induction is an extremely important scientific principle. Without it, there would be no such thing as electric generators to power our homes, businesses, and virtually everything else in our society. As important as it is, I had a difficult time teaching this vital scientific principle. Why? Because every single time I taught electromagnetic induction, I got so much sexual response from the class, they lost sight of the lesson. Now how, you might ask, could a subject like electromagnetic induction produce such "dirty" thoughts in young minds? It's because the best way to teach how magnetism can produce electricity, is to hook up a small-diameter coil of wire to a galvanometer, a device that measures electric current. Then you push and pull a bar magnet in and out of the coil of wire, pointing out that the movement of the bar magnet generates an electric current in the coil of wire each time you do so. The faster you push and pull the bar magnet through the coil, the more current it generates. I could never get through this demonstration without a lot of giggling and laughing from the class. Yes, I could show them a miniature generator with a rotating armature instead, but they wouldn't understand how it worked unless I tore it apart to show them, and then it would be useless. Even when I had other students do this demonstration, the class could never get beyond the sexual images it produced in their mind in order to learn the scientific principle.

Let me preface the following stories by saying that my only intent in telling these stories is to show how sexually saturated high schools are today, and give an insight into the mindset of the average student. Whatever attention I got from female students says much more about the atmosphere that exists in today's schools, than any quality that I personally possess. One of the more promiscuous students I knew was Kathy. She was a pretty, auburn-haired girl, with a nice figure. She was in ninth grade. She had lost her virginity shortly after her 13th birthday, and her every waking hour was spent figuring out what her next sexual exploit would be. I know all of this because she came to see me often between classes and during my planning period, and confided in me a great deal. I found out later that she even fancied herself scoring with me, and was telling the other girls about it. When I became aware of this, it scared me so bad that I called her in for a little discussion to set the record straight. She acted a little embarrassed when I told her that I had heard what other girls were saying. I told her that I would always make time available if she needed to talk, but there would never be anything other than a teacher/student relationship between us. She acted a little hurt and then made what I think was a stunning remark for a 14 year old. "I don't think you could satisfy me anyway," she said matter-of-factly. She didn't come as often after that, but at the end of the year she wrote a beautiful letter to me expressing her gratitude to me for helping her. A note to parents: The one thing I remember about the note was how she said she listened to my advice, even though she acted like she didn't. She actually was very smart, and could get grades without studying. But her passion for sex overwhelmed everything else. She sometimes would enter class by saying, "I'm so horny, today Mr. C," and wouldn't care who heard. For a long time I assumed she was an unloved product of abusive home life. She was in fact from a divorced home, but I was shocked later to find out instead that she was a spoiled daddy's girl and bragged that she could get anything from him she wanted. He always believed her, which is why she could get away with so much stuff. I was equally shocked to find out that she didn't even date! She said she wasn't allowed! When I asked her how she managed to have been with so many guys, I'll never forget her answer - "I have sex at school." She would make up excuses to stay after school, or meet someone at lunch, or even skip class to have sex. One day, I foolishly wrote her a pass, not remembering that I had written one for one of her friends earlier in the period. It was my first year of teaching and I wasn't yet wise to the ways of the deviant teenager. I found out later that while they were out of class, she was trying to get him into the restroom to have sex.

If you think Kathy's story is rare, you would be mistaken. Early one morning, we were just settling in to our first period classes when a naked female student came running across campus, and was seen by many of the other students looking out their windows. No, this wasn't the 70's, and no, she wasn't a streaker. It turns out that she was having sex with a couple of guys in one of the empty portable classrooms on our campus, when one of our teachers, Mr. Carson, walked in on them, and all three took off. The girl was in too much of a hurry to take her clothes, so she ran around campus nude until she could find an unoccupied shelter where someone could bring her clothes to her. I didn't know this particular student, by my understanding was that she was the daughter of a local police officer. I'll never forget the remark Mr. Carson made afterward. "I saw more of the student body today than I expected," he said.

Every year I taught, there would be different kids come to my classroom just to talk. Sometimes they would come by themselves, and sometimes in a group. How did we have so much time to spend together during the school day? Well, first of all, students are masters at lying their way out of class, and will play one teacher against another ("I have to go take a make-up test for Mr. X."). Most of the time, however, students are allowed to leave class because it is just easier for the teacher to let them go. If the student is a good student, they're bored and already have their work done, so many teachers allow them to roam at will. If the student is a poor student, the teacher knows they won't do any work anyway, and it's easier to just let them go, as the class may be better behaved with them gone. As for why I had so much free time, it's called a planning period, when you are supposed to be planning your lessons. Every teacher has these, and they are anywhere from one to two hours long, depending upon the schedule your school is on. And of course there is free time at lunch and immediately after school, when classes have been dismissed. Most teachers will never turn kids away when they walk in, because after all, the kids are the reason we're there. If we have to make up for it by taking work home, we will. Also, I figured that if they were in my classroom, at least they weren't getting into trouble anywhere. Anyway, when you get a reputation for being a good listener, they will come, possibly because no one listens to them at home, or because they can't tell their parents the things that they tell their teachers. Most of what I heard would cause their parents to age prematurely. I can recall four or five girls coming in one afternoon when I was trying to prepare some lessons. They were actually talking to each other more than they were talking to me. I wasn't even participating in the conversation and only half listening while preparing some lessons, but by the end of the conversation I could have told you some of the most intimate details of their lives, which they did not have to be coaxed to spill. I could have told you all of the boys they had been with including their names, and when, where, and how old they were when they "lost it," to use their language. I found, after hearing many conversations like this, that most of them "lose it" between ages 13 and 15, a few earlier, about half later.

One time when I was at my duty station between classes, I was near a stairwell which had kind of an alcove, or narrow hallway on the backside hidden from the student traffic. I thought I heard some heavy breathing while I was walking by, so I went around to the back side of the stairwell to investigate. As I turned the corner, I saw a junior boy and a freshman girl that I knew basically having sex with their clothes on. He was running his hands up and down her body underneath her clothes, from her breasts down to her hips, and she was moving in rhythm with his body, both of them in their own private dance without the music. They were both

Set Our Children Free

panting like a couple of dogs in heat. In fact, they were so focused on each other and what they were doing, that they didn't even notice me walk up just a few feet away until I startled them by telling them to knock it off. They were violating the rules, but I didn't write it up, since I knew nothing would be done anyway. There was no question in my mind that they were doing the same thing without clothes after school hours. I asked the girl later if her parents knew that she was that involved with her boyfriend, and she said no. I noticed that they broke up later that year and moved on. Should I have told her parents? You make the call.

Another girl who came by to talk frequently was a pretty blonde junior girl named Jana, who wasn't in any of my classes, although her freshman sister Jackie was. The first time I met Jana was when I was walking down a breezeway between classes. I noticed her with a couple of other girls getting some things out of her locker as I walked by. When I got to the end of the breezeway, which was only about 20 feet or so from her locker, and walked back, I noticed that she had on a different top than when I saw her a minute earlier. When I asked her why she had changed in the middle of the hallway, she said she always did because it saved time. I asked her what she would have done had I turned around at just the wrong moment, and she said, "I would have jumped you." I left it at that, but we did develop a rapport to the extent that she usually came by during my planning period. She was a "good" girl, which by the average teen definition meant that she only had one sexual partner by age 16, and only because she was "in love." Her sister

Jackie was a virgin, but that didn't even last until her sophomore year. She didn't even have a boyfriend. She just "got it over with" with a friend of hers just to avoid the drama. Jana, though, was truly put off by all of the promiscuity she saw going on. She didn't smoke or drink either. She would talk with disdain about a couple of local boys who had wild parties every night out at their house, and together had scored with about 500 girls. She had been to the parties, and had been chased around the pool by one of them, only to escape, in her mind, what would have been a rape. Like others who came by to talk, I did my best to counsel her. She was intelligent, and had some God-fearing, church-going people in her family, but she was agnostic. And I really felt her slipping away late in the year. She never returned to school and dropped out. I would ask Jackie about her occasionally but the reports got worse and worse - Jana's got a boyfriend — Jana's pregnant - Jana's on drugs. Her sister Jackie followed suit a few years later. Their story was one of my biggest regrets as a teacher. I knew that both of these girls had been brought up with sound values, albeit in a single parent household, and I did my best to help them and failed. Unfortunately, they were two of the hundreds of young people that I saw get chewed up and spit out by a corrupt culture, both inside the school and out.

My observation is that many of the students who have moral or religious objections to premarital sex, also have a great deal of self-respect, and are "A" students. Why do straight "A"s, and virginity seem to go hand in hand? It's pretty simple, really, but if you're thinking that one leads to the other, you're wrong. I have found that the desire to achieve top grades is hard-wired into a child and it is not something that can be artificially created. It means they're not impulsive, they think long-term, are willing to sacrifice present wants for future needs, and have the maturity to see the big picture. Kids who have that kind of brain think the same way about sex, and the seeds are sown early. Unfortunately everywhere these kids look, and all of the messages they receive from their friends and the media, tell them that virginity is no longer valued. Most other girls can't wait to get into a "committed" relationship so they can have sex. As long as sex is with a steady boyfriend, they're not considered "sluts" – a word girls love to call each other. Others just want to get it over with to avoid the pressure, so they basically just throw it away like it's some kind of burden or curse, and that is truly sad. I knew of one good looking football player who had girls lined up every weekend who couldn't wait to lose their virginity to him. None of them were even dating him, and he cared even less about any of them. To the girls, it was simply a rite of passage. If any sex education is needed in the schools, it is education about the real value of sex and intimacy in a marriage when two people love each other. Nobody is telling these kids how and why to say no. Federal funding for abstinence programs has also been cut, because the Obama administration doesn't really believe they work. They do.[15]

And of course, teen promiscuous behavior isn't confined to just students. Hardly a week goes by without a new report of sex between a teacher and a student, with an increasingly large number of the teachers being women.[16] Many will face no jail time due to the centuries-old double standard. Even many of the male abusers will face shockingly few consequences, and go free to continue their abuse elsewhere.[17] Many outside the school system hope that these incidents are the exception rather than the rule, but based on what I've seen, I think that is wishful thinking. The only reason some teachers don't hook up even more often with students is because of the legal and workplace restraints. Teachers risk their job at best, and jail time at worst if they give in to temptation, and the temptation is very real. Many of the male teachers I knew would frequently engage the female students in trash talk, testing the legal limit of what was allowed. This was playing with fire, because a lot of the teen girls didn't need much encouragement from a teacher to jump over the line. My daughter was the object of such sexual harassment by a teacher when she was in high school, but nothing was done when I reported it because the teacher was an active officer in the teachers' union, which backed him to the hilt even though the school had received many such reports about him.

Kids, in their minds at least, risk absolutely nothing by getting involved with teachers, which is also why they are the pursuer in many cases. There are no apparent consequences for them – if caught, even their name will not be revealed publicly – although they may boast to their friends about it. After all, they are the victim. Even when a teacher pursues the student, it is perceived as flattering by the student. Teachers are inundated with the same pop culture media sexual saturation that the kids experience every day. Combine that with what they have to see every day (See the discussion about the dress code later). Add to that the fact that a teacher is an authority figure, which, similar to police officers, coaches, bosses, and politicians, has its own subtle seduction to many students. Throw into this mix the fact that teachers and students spend many hours a day together, and you have the recipe for disaster. Without the moral, legal, and professional restraints on such behavior, the schools would resemble the worst aspects of singles clubs. Believe me, for every one teacher-student affair that makes the news, there are probably at least a half dozen that don't. Our local small town school district has had its share of teacher-student sex scandals, and the school district where I taught was no exception. It always seemed to peak around prom time for some reason, possibly because by then the teachers and students had spent an entire school year together, and there was a lot of partying and drinking afterward. Some students would choose to stay overnight, preferably at an out-of-town location, and some partied right at the prom, as evidenced by the bras and shoes the clean-ups crews would discover under the tables the next morning. Some students bragged about the teachers they had been with only to deny it when confronted. It was amazing to me that they talked about it openly in front of other teachers, with no fear that it would be reported. Most of these escapades, even when the teachers were caught, never even made the local, let alone national news, which gives you some idea of how widespread the problem is. The only time the media seems to pick up on a story of this nature, is when the teacher is a female who is very good looking.

One high school girl who lived with our family for awhile was carrying on a relatively obvious affair with a local principal, although neither would admit it if asked. She had made the same attempt with me, but met with no success. Several adults I know personally had affairs with their high school teachers a generation ago. My own adopted daughter was likely the product of an affair between a teacher and a 17 year old student during that same time period. Long before a parade of women teachers started making the national news for having affairs with male students, we had several female teachers quietly fired in our local school district for the same offense. One teacher would have gotten away with it, except that a student took a picture of her with some other students at an off-campus party with a beer in her hand. This particular teacher reportedly picked a special student every year and gave them names coincident with that school year, such as, Mr. 2002, Mr. 2003, etc. It goes on way more often than most people want to believe. I have seen studies on the percentage of students who have had sex with a teacher, and it was in double digits. I don't know how accurate the data was, but given my experience, I am not surprised.

In my case, I had the self-discipline of moral and spiritual restraints in my life. Absent that, many teachers have only the fear of prosecution or termination to hold them back. For most of those teachers without an internal value system, the artificial restraints of job and/or criminal sanctions are not always effective to deter them from stepping over the line. John Adams, one of our nation's Founders, put it succinctly when he said, "[W]e have no government armed with power capable of contending with human passions unbridled by morality and religion ..."[18] The lack of morality in our society is reflected vividly in our schools. If your high school or middle school child has a "crush" on a particular teacher, I would suggest that you have a talk with both of them, and then monitor their cell phone calls, text messages, face book pages, etc. Don't assume it's innocent puppy love.

When the student is the pursuer/harasser, the teacher has two choices: ignore it or give in. As I said earlier, joking with students about sexual remarks is just playing with fire, and akin to giving in. I've repeatedly had to ignore remarks made to me by female students as I was trying to conduct class. Despite the fact that I would discourage such remarks whenever they came up, I still got them. I remember one girl named Carla embarrassing me in front of the class as I was passing out papers, by remarking how much she liked this or that body part, and some other trait about me. After each comment, I told her that it was inappropriate for classroom conversation, but she continued, and I continued to ignore her. I once had a cute, dark haired sophomore student named Laura in my class, who was so aggressive with her remarks that some of the other students would tell her to knock it off and suggest that she get a motel with me. She wasn't sleazy about it, and wasn't a promiscuous girl, but she made it clear that she liked me, and that she was available should I ever be so inclined. She had only been with one guy by that time, and they broke up because he insisted that they have sex every single day, and got mad at her if they didn't. She told me all this when she came in to talk, and I spent time off and on the next two years counseling her, and helping her with her assignments from other classes. When she graduated, she wrote me a beautiful card expressing her gratitude, which was a far greater reward than any short term fling I could have ever had. Another time I remember running into Patti, our reigning prom queen, one day as I was walking across campus. I hadn't seen her since she had been won the title a few weeks earlier and I congratulated her, and told her she deserved it. I was walking away from her at the time, but my words stopped her dead in her tracks. She gave me a huge smile and stared at me with those big beautiful eyes, and expressed her thanks. The "thank you" was normal, but the stare was not. What I saw in that stare was not what I expected to see in the eyes of a student who, to my knowledge, had a good reputation. The stare told me more than I wanted to know, and whether my perception was right or wrong, I felt disappointed as I hurried away that she did not seem so innocent anymore.

Speaking of the prom, I once had a girl in my class ask me to the prom. I am not making this up. Sara was a beautiful girl who sat in the front row and wore a lot of stylish outfits. She had the looks of the classic sexy blonde, but didn't seem to know how pretty she really was. She actually seemed to have a low self-esteem for whatever reason, so I would sometimes compliment her on what she wore to cheer her up. This was something a teacher had to do very carefully or they could be perceived as having the wrong motive. I would compliment the outfit, not the person, as in, "That looks nice," as opposed to, "You look nice." I didn't know much about her personal life other than the fact that she didn't have a current boyfriend. She and Becky, another girl in the class, asked to take a picture with me one day, so I complied thinking it would do no harm. They gave me a copy of it later and I still have it. One day she caught me completely by surprise and asked me if I would take her to the prom. I thought she was kidding but she wasn't. I politely pointed out that teachers weren't allowed to date students, assuming she knew that. She then shocked me by saying that one of our teachers took a girl to the prom last year, mentioning both the teacher and the girl's name. I had a hard time believing this so I asked this particular teacher about it later and he deflected the whole conversation by generally deflecting the possibility, but he never specifically denied the allegation. To this day, I don't know whether there was some secret arrangement between this teacher and student involving the prom, or not. As for Sara, I felt sorry for her because she was the type of girl that any guy should have been proud to be with, and here it was a couple of weeks before the prom and she had no date. And even though I knew she had somewhat of a crush on me, she must have really felt desperate for a date to ask a teacher. I felt bad enough that I told her I would take her just as a courtesy if it was allowed, but I couldn't. She seemed genuinely disappointed. What really amazed me about this incident was that Sara really thought it would be OK for us to go together. She wasn't stupid, and I'm sure she was aware of the spoken and unspoken rules against such behavior, and yet it didn't deter her. The only explanation I have is that, in the eyes of many students, inappropriate teacher/pupil relationships are the expected norm and no big deal.

Sometimes the attraction wasn't obvious but it was still there. I remember a girl named Bobbie who I had in my class during her senior year who gave me grief all the time. It wasn't grief in the bad sense, but she was always putting me down in a friendly cutting-up sort of way, but with a definite attitude. I wondered the entire year what I had done to make her feel that way, but since she wasn't disrespectful, I pretty much ignored it. A year later I ran into her brother who had graduated a year or two ahead of her, and he told me out of the clear blue how wild she was, and by the way, how much she liked me, even though I never realized it. Another girl I would put in this category was Lana. She and another girl I coached sat next to each other in my classes for two or three years. I remember one year explaining how mass affects the gravitational formula. The greater the mass, the greater the gravitational force, not just for planetary size objects, but for relatively small objects like ourselves. The greater the gravitational force, the more two objects attract one another. To emphasize the point humorously in a way that they would remember, I told all of the girls in class one day, tongue in cheek, not to worry about putting on a few extra pounds over the Christmas holiday since it would just add to their mass and make them more "attractive," gravitationally. So the last day of school before Christmas, both Lana and her friend wrote me a card telling me how they planned to eat a lot over the Christmas holiday so they would be more attractive to me. Only a few weeks before she graduated, Lana came by and apologized to me for talking too much in class, an incident for which I had to write her up. In apologizing, she was in tears and recalling how we had "been together" for three years now and what a shame it would be to end it this way. I told her it was OK and wanted to hug her since she was crying and I felt so bad. There really was some genuine affection between us, but hugging was something you just didn't do with students, and there were only a couple of unwritten exceptions – as a coach on the athletic field after a big win, and at graduation. I never saw either student after they graduated, although they both went to good schools and I expect they did quite well since they were top students and came from excellent families.

Sometimes the affection, or pretended affection, wasn't subtle at all, and was the result of an ulterior motive, and I don't mean sex. Some girls try to manipulate the teacher to get a favor or a grade. In their mind, it's easier to cozy up than to put in the work. They would approach my desk and ask if I could give them more time to do their assignment, raise their grade, write them a pass, or any one of a number of other privileges. The request was always accompanied by a big smile, a tilt of the head, a syrupy sexy tone in their voice, and a body language that said, "I'm cute and you can't resist this request." Sometimes I would ask them straight up, "Does that work with other teachers?" Most of the time they would immediately drop the façade and state honestly, "Yeah, pretty much!" They would always seem so crushed when they realized that their feminine charm wasn't working. I could only conclude that they were very aware of their sexuality, and that it really did work on some of the other teachers. The play-acting, however, really came to a screeching halt near the end of a quarter or semester, when grades were due to be sent home. This was panic time. "I can't go home with that grade, my parents will kill me," or even more likely was, "I'll lose my car," or "I'll be on restriction." I can still hear the whining and crying years later. They would do anything at that moment to avoid the consequences. The resulting negotiations that took place would make the State Department blush. Forget the fact that they hadn't done a thing all semester; it was now **Judgment Day**. Once these kinds of girls found out I wouldn't compromise or play favorites, their attitude was usually cool towards me the remainder of the year. This despite the fact that I would tell all of the students at the beginning of the year that I held one principle above all others – you get the grade you earn. Personal feelings did not ever influence my grading. I told them that I had sometimes failed kids I really liked, and had given A's to kids I very much disliked. I also noticed that I never had any of the boys in my class give me the "you-can't-give-me-that-grade-or-my-life-is-over" routine. You can draw your own conclusions from that, but I suppose it's because they didn't try to fool themselves and were more willing to accept the consequences of their behavior than the girls were.

There were a few students that I developed a close relationship with during my teaching tenure, although I wouldn't call them friends. I don't believe, except in very rare cases, that true friendship can develop between teacher and student for many reasons, most of them obvious. But there can be closeness between a mentor and a life-learner when there is mutual respect and a deep caring for the well-being of the other. Two such students were Angie and Megan, who were friends with each other, and took one of my classes as juniors. Amazingly enough they elected to also take another one of my senior classes together the next year, even though they knew they couldn't manipulate their way to a good grade. Of course they didn't need to, because they were smart enough to get good grades. Notice I didn't say they were good students – just that they were smart enough to get B's without putting in much work. And did I mention that they were both drop-dead gorgeous. I'm sorry, but just because I never got sexually involved with my students, it doesn't mean that I'm blind. They confided in me a lot (trust me, I don't ask for these details, but girls are more than willing to share). Megan had been in a sexual relationship with her boyfriend since her junior year. I warned her about the consequences, but sure enough she got pregnant, or at least thought she was. She thought her world had ended, because at that time she valued her education and saw herself going places. She was in a panic and crying when she told me. I immediately walked her outside of the science building and prayed with her (Yes, prayed. So arrest me). She felt better and told me within a week, I believe, that it was a false alarm and she wasn't pregnant. I have always hoped that she told me the truth and that she didn't have an abortion. Angie had three sexual partners by the time she was a senior, but wasn't dating anyone in particular at that time. Sometime during their senior year, they got friendly enough that they wanted me to start doing some of the sharing – such as the intimate details of my marriage. "C'mon Mr. C," they said, "we're 18 now, we understand." Of course, I declined. That would have definitely crossed the line. There is no question in my mind, however, that either girl would have slept with me if that had been my motive.

One of them said as much to me her last day of school before she graduated, admitting her feelings and attraction for me. I don't say that to boast, but just to illustrate how easy it is for teachers and students to get too involved with one another. If it could have happened to me, and trust me I'm no Mr. America, it could happen to any teacher. At the beginning of the second semester of their senior year, Megan started missing classes. When I inquired, Angie told me that Megan had taken a job. I thought that maybe she was missing class because she was just tired from her job, but pretty soon she wasn't coming to class at all. It turns out that she had gotten a job as a waitress at Hooters, the controversial national chain restaurant known for their skimpy waitress outfits. I phoned home and tried to get her mom and friends to persuade her to finish school and get her degree, especially since she had academic potential, and only one semester left. But she never returned and I never talked with her again. This was a big disappointment for me as I thought she had a good head on her shoulders. I heard a couple of years later that she had a boyfriend, a baby, and possibly a drug problem. Angie did graduate, and I talked with her once since then, and although she was doing all right, she wasn't going to school or doing much of anything else. I would put both of these girls in the category of great potential wasted, and two of the biggest disappointments I had as a teacher. This seemed to be another continuing theme I observed throughout my brief teaching career; that is, if college or any other major goal isn't in the picture, kids tend to just float downstream and bump into whatever comes along. With no long-term goals, the "bump" unfortunately becomes a permanent attachment. Kids with

Set Our Children Free

long-term goals swim upstream, and are not deterred by a few bumps here and there.

An example of how you can get in trouble even when you mean well was when I wrote up a girl named Heather for a dress code violation. She was a beautiful girl, even as a freshman, but she came to class one day wearing a provocative T-shirt that said, "Save a horse, ride a cowboy." The shirt included a picture of a cowboy with his shirt off, complete with the requisite six pack abs, riding a horse. When I told her the T-shirt was a dress code violation, she said that the phrase on the shirt was a popular song, so in her mind I guess that made it OK. And, as I had heard way too many times before, no other teacher or administrator had said anything about her shirt all day long. This was an overused attempt many students use to make you feel like you're the only one who is interpreting anything bad about their attire, and thus make you feel like a pervert. Anyway, by the letter of the rule, her shirt was a violation, so she had to go to the office. The mistake I made, however, was actually caring about what message her shirt was sending to the male population of the school, since she was only a freshman. Before she left class, I gently reminded her that boys could possibly misinterpret her intentions by how she presents herself, such as wearing that type of T-shirt. Well, that got me into a parent-teacher conference pretty quick. You would think that any parent would appreciate someone caring enough about their freshman daughter to want to protect her, but that was not the case. Her mother was angry at me, and said she felt very uncomfortable about what I said. Since the school principal normally caved in to the parents, I didn't like the way the conference was going when one of the school counselors, Ms. Wainwright, spoke up. Now, I didn't usually have much confidence in the school counselors, but I was pleasantly surprised that Ms. Wainwright took my side. She took over and lectured the girl like a preacher, with pretty much the same sermon I had given her. When it was over, the parents were mollified, although still not completely satisfied, since they hadn't extracted their pound of flesh from me. I ran into this mom outside school a few years later, and I could tell she was still steamed at me. To this day, I don't know why, especially if the daughter told her the correct version of the story.

When high school girls get dressed in the morning, modesty is not on the minds of most of them. Yes, some of them complain that they can't buy modest clothes, and that is partly true, if they want to dress hip. By definition, most of the clothes being sold to that age group are immodest by design. So if a girl has the right figure and wants to look cool, she will have little choice as to what to wear. That is a function of our culture, a culture which also infects clothes designers. There are many girls, however, who do dress modestly, but when they do so they are purposely making themselves less attractive, at least on the outside, to the male students. That is an uphill battle mentally and emotionally to the average high school teen girl. She can choose to wear the attention-getting tight tank top or blouse, with a short skirt or short shorts, or she can choose the casual unassuming blue jeans and loose T-shirt. What do you think she's going to choose when she will spend all day parading in front of cute, testosterone driven boys? This is one reason, among others, why school uniforms may be a good idea. It removes a great deal of the pressure, as well as developing an atmosphere of equality, unity, and teamwork, without the unnecessary distractions.

Of course our school district had a dress code, but it was largely ignored because the assistant principal, Mr. Green, made it clear that he didn't want to enforce it. He never gave his reasons, so I'll let you speculate why, but it was a huge mistake. It was already intimidating for a male teacher to send a girl student to the office for a dress code violation because he would be accused of leering at the girl and being a pervert, and the boys in class would accuse the teacher of wanting her. This was a very frequent occurrence. But you had to draw the line somewhere or it would get completely out of hand. As it was, you had to teach class while trying not to observe and remember the color and pattern of the panties you observed from the front of the room. If you went to the back of the room or walked down the aisles it wasn't much better, as rear ends and thong underwear forced themselves onto your visual senses. Even when addressing girls who were standing, you had to concentrate on maintaining eye contact instead of observing the bosoms that were ready to fall out of their top. And jeans and shorts weren't safe either, since the girls would pull up their way-too-short top and play with their belly button jewelry on their toned and tanned midsections during class, and invite you to comment on their bling-bling, which of course, you couldn't. If you did send them to the office because their skirt or shorts were too short (the standard was finger tip length with arms at their side), they would hunch their shoulders so that their fingertips fell somewhere just barely south of their waist. The tank tops, or any other tops, were supposed to cover their waist in any position, sitting, standing, or with arms raised, but it was never enforced. Most teachers just gave up. The boys weren't much better, but their dress code violations were easier to enforce, because most of the violations were for obscene T-shirts, ball caps, or pants too low to cover their butt. I remember teaching class one day and interrupting myself in the middle of my lesson to read a ball cap on a boy's head in the front row. It said, "It's not going to lick itself." Once the hat got my attention, he smiled because he knew he was busted, but he did mention that he had worn the hat all day and no other teacher had said anything to him. I wish I had a dollar for every time I heard that, but I simply could not tolerate what I felt was wrong. I could never remember all of the illegal T-shirts the kids wore, but I recall one that said "Job openings: many positions available," which sounds innocent until you notice that the stick figure drawings on the shirt are having sex in different positions. Others contained offensive language or a double entendre', which the kids would argue only contained the innocent meaning, but most bragged about sex in one way or another. Most of the girls' illegal T-shirts contained come-ons or flaunted their attributes, such as "Stop staring at my flags," which displayed a rebel flag on each breast, or "Objects underneath shirt are larger than they appear." So with the dress code unenforced, what you basically had, was not a learning institution, but seemingly a super-saturated hormonal campus full of street walkers displaying their wares to an equal number of johns bragging about their sexual prowess.

Set Our Children Free

And if the dress code was bad during the school day, it really went out the window after school, during any kind of athletic team practice. Short shorts and tank tops rolled up to expose the midsection were everywhere. School related activities weren't any better. One of the girls' teams I coached had a car wash every year to raise money for uniforms and accessories. They always started off innocently enough when the girls showed up in the cool of the morning. As the weather warmed up during the day, the clothes began to come off, exposing more flesh. This, of course, attracted more customers, which in turn meant more work washing cars, which necessitated more disrobing until most of them were down to the barest minimum top and bottom. The girls loved the attention they would get from the customers, most of them guys, who cared less about getting their car clean than getting an eyeful. One of the jobs the girls liked most was standing out on the street "advertising" the car wash by holding up signs. I got the impression that they enjoyed advertising themselves even more. Every other car would honk at them, shout something, or just stare. It was amazing to me how the girls were so aware of their sexuality. Many of them are as sexually experienced as their college counterparts were a generation ago. Interestingly enough, the private school where I taught my final year had a very strict dress code, and it was refreshing. It made for a much better learning atmosphere. However, once school was over and the athletic practices started, there was little difference from the public school, as the girls roamed the small campus in various stages of undress. It didn't make any sense to me since it seemed to defeat the purpose of the dress code to begin with. The only difference was the time of day.

One time, I was performing duty just outside the science building between classes. "Duty" is a polite way of saying that you hang around a designated area and watch the students to make sure that the number of felonies taking place within your area is kept to a minimum. As I turned to look behind me, I saw nothing but long bare legs and a rear end. No upper torso, just the naked lower torso (well, she was wearing thong underwear) of a girl bent over facing away from me as if she was either giving me a moon shot, or auditioning for a strip club. It was neither. Some boy had pulled her skirt down to her ankles and run off. I had just happened to look over at the exact moment she was bending over and pulling her skirt back up. I walked over to find out what was going on, and there was a look of fear and embarrassment on her face. When she saw me, she thought that she might be in trouble, but several of the students nearby backed up her story. Technically, this could have meant a sexual battery charge, but absolutely nothing was ever done after I reported it, first to the assistant principal, then to the principal, and then to the superintendent and school board. No discipline referral, no suspension, no criminal complaint. Nothing. Why nothing was done is a subject which I discuss further in Chapter 4, but suffice it to say that school discipline enforcement was as lax as dress code enforcement.

One of the teachers in our local school district got fired for viewing porn on his computer. One of his students, a 15 year old named Shelley, who was very close to our family and attended that particular school, wasn't even fazed by the allegations. Her only lament was that he hadn't hit on <u>her</u>

yet. When I asked her why she would welcome advances from someone like that, she was honest about how flattered she would be because he was good looking. And this was from a straight A student and a virgin, not a troubled teen who needed love. She would go on to do very well academically in college, although by age 17 she was dating a 28 year old. This was another continuing theme I noticed – the number of high school girls who date men in their twenties is staggering. The more mature the girl, the more likely it was to happen, presumably because they found high school guys either too immature or too broke for them. I told her I understood why she would feel flattered that a 28 year old would find her attractive, but, no offense, what would a 28 year old have in common with a high school student other than biology? She eventually broke up with him because she said he was too immature. I wanted to say, "Really? What tipped you off?" I distinctly recall when I was a senior in high school refusing to date a good looking ninth grader who wanted to go out with me because I felt she was too young. And when I was in college, it was considered taboo and "robbing the cradle" if you dated a high school girl more than a year younger than you because of the differences in maturity and culture between high school and college. Now, it is commonplace for high school girls to date guys in their twenties, and do it with full approval of their parents, which I don't understand. Similarly, senior boy – freshman girl dating is not even blinked at. I specifically recall one senior boy bragging that he made a point of targeting freshman girls because he knew how vulnerable and impressionable they were, especially if they were being hit on by a senior guy.

I remember a girlfriend/boyfriend duo I had in my class one year whom I'll call Wanda and Mike. She was smart and could get grades without trying, and she was a very stylish dresser. Mike was just the opposite – no slave to fashion – but a real likeable kid. He was a lot slower than her, intelligence wise, so I let them sit together because they were quiet and she helped him with his assignments. She would always talk to me about making sure he kept up with his work, acting more like a mother at times than a girlfriend. I thought it was great that she had such a concern about his academic welfare. Near the end of their senior year I noticed that they had broke up. When I asked him why, he said that she had cheated on him with a much older twenty-something guy, who I recall she subsequently moved in with. This shocked me as she seemed so sensitive and compassionate towards Mike. She definitely was way more sophisticated and "older" acting than him, but I did think that she genuinely cared about him. Although he seemed to take it well, without her guidance, he missed a few more classes than he should have, and therefore was not allowed to graduate with his class. I felt both sorry and worried for both of them.

Even though many of the high school girls have no qualms about hooking up with older men, I was still surprised by an incident I had with one of my college students, where I taught night classes part time as an adjunct. A very attractive 19 year-old student named Amy had stopped by after class to talk about a problem one evening early in the semester. The topic of conversation wasn't academics, and I don't remember much about it, but I listened and gave her some advice. I got no indication from her at that time that she was attracted to me in the slightest. After the next class, she followed me out to the parking lot and called to me from behind as I was approaching my car. She told me that she had dreamed about me the previous night, and that in the dream we were really dating. I thought that was a strange way to open a conversation, but I made some reply about how dreams are just dreams and not reality, but this did not deter her. She continued telling me about all of the things she liked about me including my ethnicity (Italian), my moustache, my rear end, etc. Once again, I made some dismissive statement, but not to be discouraged, she continued telling me how, upon waking from her dream, she was so aroused that she just had to "take care of business." I just looked at her in disbelief, and said, "Amy, I can't believe you just told me that." Smiling, she said, "I know, I guess I'm just a troublemaker." This was perhaps the biggest test of my commitment to be a professional and a role model, not to mention a faithful husband. It was the first time I was tested by a female student when no artificial restraints existed to stop me. I wouldn't go to jail, and probably not even lose my job if I gave in. For most guys this would have literally been a no-brainer, i.e. the brain tends to disconnect at times like these. Was it easy to resist the temptation? Absolutely not, but the right way is rarely the easy way. I believe we will all account someday for everything we have done, and this is one of the problems with public education today – little to no accountability on either an academic, behavioral, or personal level. This reflects the trend our society has taken. Unfortunately, fewer restrictions and less accountability in personal behavior doesn't produce more freedom, only anarchy. And absolute freedom won't liberate at all, but will bring bondage, on both a personal and national level. If that's not true, then why do so many Hollywood celebrities, who can live any lifestyle they want and indulge their every fantasy, find themselves on psychiatrist's couches and in drug rehab centers? In case you were wondering; no, I didn't sleep with Amy. There were several other female students at that same college, some who were married, that made their affections known to me, but none as brazen as Amy. One student named Tracy used to sit in the front row, wink at me, and make other not-so-subtle remarks during breaks or while I was working with her on a math problem. Resisting temptation with her was a little easier, however, since she had a tattoo. I'm sorry, call it a fault or a prejudice, but I just think a tattoo, any tattoo, ruins the natural God-given beauty of the female form.

It seems that once teen kids become sexually active, it is fruitless to talk with them. They have lost something very precious in my opinion – their innocence. And they don't seem to ever be able to recover, short of a life-changing experience. Even after they're married, they carry a lot of baggage into that relationship and the outcome is rarely good. Based upon my experience with hundreds of young people I've counseled before, during, and after my teaching tenure, kids who wait until marriage to have sex, have better and longer lasting marriages than those who don't. In fact, there have been several reliable studies confirming that divorce is much more likely for couples who have lived together before the marriage.[19] And contrary to the popular belief that experience is necessary, sexual satisfaction during marriage is much higher among couples who are monogamous, and among those who didn't sleep together before the marriage.[20] That's just some of the reasons, among others (STD's, pregnancy, emotional harm, much higher rates of cervical cancer, etc.) that kids should not only be taught abstinence, but be given good reasons <u>why</u> they should wait.[21] Abstinence-only programs have been proven to work,[22] and in my opinion, would be even more effective if taught in the proper context, and by the right role models. The value of sex is that it is much more than just a physical act – it is an intensely intimate and emotional connection between two souls, not just two bodies. Those who abuse this precious gift just for a little entertainment, pay a terrible price sooner or later, contrary to the sex-without-consequences picture that Hollywood inundates us with on a daily basis. Unfortunately, they are not the only ones to pay, as the rest of us in society have to pick up the pieces from the shattered lives, single moms, divorce, disease, and emotionally scarred kids. I challenge any teacher to deny what I'm about to say. My estimate is that about 75% of all my below average kids were products of broken homes. On the other hand, divorce was rare in the homes of the top performing kids. How do I know this? In addition to my personal conversations with these kids, the evidence is all there in the parent contact information in the school office. The lower performers almost invariably had two parents/guardians with different last names, or more likely just one parent listed, or a grandparent. Conversely, the top kids almost always had two parents listed with the same last name as their own. Coincidence? I don't think so. Many studies have shown such a direct link between academic behavior and divorce, even after controlling for IQ and socioeconomic status.[23] None of this would come as a surprise to any teacher I knew. If it was only academic performance that suffered, it would be tragic enough, but it was easy to see from my point of view that the emotional baggage these kids carry manifests itself in every area of their behavior. In fact, kids from disrupted or divorced households miss way more school, and are much more likely to have been suspended, expelled, or dropped out.[24] I maintain that a student's home situation is not only the most consistent predictor of success, but it is the **greatest single influence** on their life in and out of school.

Many of our high school girls are not just dating older men, they are having babies with them. Contrary to the opinion of those who would push for dispensing birth control at school, studies have shown that the majority of high school girls who get pregnant do so with older, twenty or thirty-something men.[25] Yes, you heard right! The big lie is that all these teen girls are getting pregnant because irresponsible kids are fooling around, and can't help themselves, necessitating a call for more "comprehensive" sex education in the schools. You may wonder then why there aren't more prosecutions against these older men for statutory rape. A lot of people wonder the same thing, as parents and prosecutors are reluctant to press charges. Instead, our educational establishment's answer is handing out condoms, and assisting unwed mothers with abortions and nursery services. We allow the girls to continue to attend school throughout their pregnancy, and we set up a nursery to take care of their kids during school hours. Our nursery was at the very center of our campus, with an outdoor playground. What do you think all of this teaches other girls who are considering being sexually active? Contrary to the stigma of a generation ago, our schools and our government now rush to the aid of any unmarried pregnant female, covering them with money and benefits. Want to get food stamps, ADC, WIC, social security, housing, Medicaid, and financial aid for college? Get pregnant - just don't get married. Want to get out of your parents' house and be financially able to live "on your own?" Get pregnant - just don't get married. Yet the promises of government are financial fool's gold, because statistically, the fastest and surest way to poverty is to have a child out of wedlock. In fact, children living with single moms are five times more likely to live in poverty compared to those living in a married household.[26] We have rewarded promiscuity and bad decision- making like no other nation, even though the consequences of these unwed pregnancies take an enormous toll on our society. An entire welfare state and criminal culture has been spawned by this kind of irresponsible behavior. And yet we are afraid to teach abstinence in schools? The best we can do is have sexually active girls carry around baby dolls, who cry and wet their pants at certain intervals to educate girls on how much work it is to raise a real child? Yes, we really did this at our school. Wherever "comprehensive" sex education has been tried, it has not worked,[27] and in many cases resulted in more, not less, sexual activity among high school kids.[28] Why? It's because laws and rules serve as more than just a restraining force. The law is a teacher. The law teaches values by rewarding some behaviors and punishing others. If you hand out condoms and birth control, whether you are a parent or a school, you have given tacit approval to the kind of behavior that requires teens to use them. Kids look to their parents and schools as authority figures. They need these authorities to tell them "NO," out of love, and give them good reasons why. Even when these authorities tell them it's not OK to play, but if you do, be careful how you play, the message is: "We understand, you can't help yourself...wink, wink." Such a message by an authority figure is a green light, not a caution light. And don't expect that school officials will be honest with you about the content of their sex education curriculum. We got a taste of this when our daughter was a freshman in high school. She was forced to watch, in a coed classroom, a video of an adult male performing a testicular examination on himself! When we questioned the value of showing freshmen girls a testicular exam, the school and health department officials (who showed the video), with a straight face, replied that the girls could go home and explain the technique to their fathers! They never apologized or admitted that showing the video was unacceptable. If you think this kind of behavior is abnormal for public school officials, you would be wrong. I have seen some of the literature given only to school administrators by organizations that produce such controversial programs, and found instructions on how to hide the program's content from the public, and how to blunt public opposition when it does happen. Think about that for a minute. We pay for these programs with our tax money, and they don't want us to know the content because we might disagree with what they want to teach our kids. Maybe it's because polls show that the overwhelming majority of parents and teens want abstinence taught in the schools, not comprehensive sex education.[29] Who do these schools and kids belong to anyway? Obviously, their thought process is that once you enroll your children in school, the school owns them.

Abstinence programs work,[30] particularly if kids are given good reasons why they should remain abstinent. There are plenty of good psychological, physiological, and sociological reasons, even absent the religious and moral issues, for a teenager to remain abstinent until they are mature enough to take on marital responsibilities. And these programs would be even more effective if they were administered by school officials who actually believed what they were teaching. Although I never taught these programs in the schools, I recall sitting in on more than one session which another teacher taught. In such classes, I already knew the first question that would be asked of the teacher, and sure enough it was: "Did _you_ have sex before _you_ got married?" Once the teacher answered "yes," there was no point in teaching the remainder of the class. Kids know hypocrisy when they see it. I distinctly remember the one class where I heard another answer, and it was at Hillside Academy, the private school where I taught after I left Hancock. The teacher, Mr. Davis, was a coach who was well respected, and the class was segregated, i.e. boys only – which is the proper way to teach physical education for many reasons. Mr. Davis, who had been married long enough to have teenagers himself, said in reply to the above question, "No, my wife is the only woman I've ever been with." There was no laughing or insults from the boys, like you would imagine, only quiet respect. I believe he taught the boys more about sex education with that one statement than he ever could have with a semester worth of "comprehensive" sex education.

The truth is, we expect too little from our teenagers, both academically and behaviorally. One of the best relationships I developed in my years of teaching was with a girl named Tonya. She was a championship athlete who earned a full athletic scholarship, but could have gotten one for her academic record as well. I taught her in Physics and coached her on one of my teams, so we had ample opportunity to talk. She was a virgin, and unlike many girls, was proud of it, even though she never attributed it to a strong religious faith. But in her mature way of thinking, she simply could not understand how promiscuous many of her classmates were. She was old fashioned enough to believe that sex was special and was to be reserved for a special relationship. In other words, she respected herself and her body too much to engage in such behavior. She had a boyfriend now and then, (college age of course), but nothing serious, and these relationships never consumed her life like it did with most other girls. She knew what her goals were, and focused on them, and was willing to delay gratification to achieve them. However, even she felt the pressure to conform. I knew this, and knew that she was continually hearing the message that everyone was doing it, or at least lying about it. So I found a legitimate study on teen sexual behavior and gave it to her. I don't remember what the actual statistics were, but she was amazed, as I was, at the percentage of girls that were still virgins even after age 20, although by that time they were a significant minority. The article included testimonials of girls who had waited, the reasons why they did, and that they were happy with their decision. The study encouraged her a great deal and she thanked me many times. Her graduation was one of the hardest good-byes I've ever experienced. I lost track of her at the end of the ceremony because there were so many people there, but she managed to search me out and find me, and we hugged for a long time. She gave me a beautiful thank-you card with her phone number and told me to keep in touch. I've only talked to her twice since, but she came from a good family with good values and I knew she would do well. She graduated college in three years, earning All-American honors in her sport, while considering a master's degree. To this day, I miss her. Time and time again, when high school teens are honest, they will tell you that what they need is not more sex education, but someone to tell them how to say "no."[31] Why aren't we listening? Most who say "yes" end up regretting their decision.[32]

I know that most of this chapter sounds negative, but I would be remiss if I didn't mention some other students, like Tonya, whose morality ran counter to the trend. They were a joy to teach, and just talking with them was refreshing and a source of hope. They typically gravitated towards the leadership positions in the student body, because they had a high degree of character. If they had been followers, they would be acting just like everyone else. More often than not they came from a conservative/religious background where they lived with both original parents, and the parents were middle class, but not always. One year I taught a beautiful couple I'll call Hank and Tiffany. He was a football player, and she had prom queen looks, although I don't remember her winning any of the pageants. It was a fairy tale romance. They had been girlfriend and boyfriend for several years, and had the kind of relationship any parent would want for their child. Before I had ever talked in depth with either one of them, it was obvious that their relationship wasn't based on sex. And because it wasn't, there was no drama, no fighting, no embarrassing behavior, no hanging on each other every chance they got. There was just a quiet, playful contentedness about them as they held hands and talked. And even though they sat next to each other in class, they had so much self- discipline that they never talked to each other during instructional time. They understood that it was all about learning, and that the time for talking was later. They were seniors and would be attending college together, she told me, but they would probably get married sometime before they graduated. They had no intention of sleeping together until the marriage took place. After checking with them a few years later, I'm happy to report that they had both earned their degree ahead of time, and were doing splendidly well in their chosen field. They have also married and are blissfully happy. They've also done mission trips together to aid poor children overseas. None of this was any surprise to me, as it was the easiest prediction I ever made.

Another such girl I spent a lot of time with was Charity, who was not in my class but on one of the teams I coached. She didn't need to tell me she was a virgin with high moral principles - it was obvious, and again it was because of her Christian faith. And contrary to what today's culture would have young people believe, girls like Charity aren't virgins because they're unattractive or can't get a boyfriend. Quite the opposite. Charity was the reigning beauty queen from a nearby county prior to transferring to our school. She was always happy and upbeat, and as beautiful on the inside as she was on the outside. As with Tonya earlier, we developed a close caring relationship that was nothing but clean and pure. Two other examples, were a couple of guys I'll call David and Doug. Doug was a top student who was in my Physics class, and whom I coached in two different sports. He was good looking and very popular. He had a nice girlfriend, but could have had any of the girls on campus had he so desired. In fact he was so popular, even as a sophomore on my basketball team, that I heard way more cheers from the coeds when he scored a basket than for any other player. He was a great student and had enough college credits to have graduated with a two-year AA degree from the local community college by the time he graduated from high school. He was of very high character, and now that he has his college degree, he is fulfilling his goal to be a coach. I run into him now and then, and he hasn't changed. He was our high school's version of Tim Tebow that you never heard about. David was another student who was a good athlete, and whom I also coached in more than one sport. I had also coached his older sister a couple of years earlier. He was of equally high character, good looking, and could have dated most any girl on campus, but chose not to. His parents were great people, and spending time with them gave me some insight into why David was the kind of kid he was. They kept him grounded and he was definitely not spoiled. He didn't even have a car, and carried a Bible everywhere he went, right along with his books. I remember one time at an away game, when we stopped to eat at a Burger King on the way home, he had it figured to the exact penny how to order from the menu so that he could get the most food for the least amount of money because he never had an unlimited amount of cash. If he put that much thought into buying a value meal, is it any wonder that he was successful in everything else? He would go on to graduate from a prestigious university and get married. He is now in law school, where we definitely could use such people of character. He was valedictorian his senior year and spoke unashamedly of his Christian faith during his valedictory speech. He received a long, enthusiastic ovation. And although he attributed his success to his relationship with Jesus Christ, there was no selective censorship of his speech by school officials - a far too common occurrence these days. I find it ironic that many school officials edit out such religious references, when the role-model student delivering such a speech could probably do more good by sharing the faith that made them successful, than any other program or class that the school had to offer. It is also disgustingly hypocritical to censor a student in an environment that, according to these same officials, is supposed to celebrate freedom of thought and expression.

CHAPTER 4

BEHAVIOR AND DISCIPLINE

There he was, a student whose name I'm happy to forget, but I'll call him Erkel. He was strolling, or should I say dancing his way down the hallway outside my classroom visible to me through the glass door. The reason for the spring in his step was the music in his ears. One problem. He was breaking the school rules by having a portable CD player during school hours. The rule had been implemented during the past year, and teachers overwhelmingly applauded it because they knew that having CD players, i-pods, and the like during school hours was a definite detriment to the learning environment. Students, of course, widely ignored the rule, and yet they were shocked, yes, shocked I tell you, when these devices were taken from them. They were kept in the office until their parents came and picked them up. In fact, I had just taken one of these portable brain-blasters from Erkel just a few days earlier when he was openly wearing it at lunch outside the science building. He was very disrespectful to me at the time as if I was abridging one of his Constitutional rights. Therefore, when I saw him walk past my room only a few days later with the same or similar device, basically flaunting the rule violation, I interrupted what I was doing, stuck my head out the door, and told him to give it up.

Instead of simply complying because he had been caught red handed, he interrupted what I was saying with loud yelling, and without looking back, quickened his pace so he could make it to the sanctuary of his classroom. Although this was totally unacceptable behavior, it was also very normal for many high school students because they know there are few if any consequences for breaking the rules and being disrespectful to, or ignoring a teacher. I was never able to accept this attitude from a student and I never understood why any of the other teachers did either. Erkel then disappeared into the classroom next door and I followed him. This happened to be Ms. Doolittle's classroom, who in my opinion, coddled her students excessively. This incident only reinforced that belief in my mind as he seemed to have no fear of wearing the CD player into her room. I opened the door to her classroom, and Erkel began crying out to Ms. Doolittle to save him from the mean old teacher that was picking on him. I politely asked Ms. Doolittle if she would please take the device from him. Her expression said it all: "How dare you interrupt my class with something as trivial as a school rule violation!" After a frown that looked like she swallowed some vinegar, she asked Erkel to give up his player. It was clear to me that she didn't want to do it, but she did so, reluctantly. He ignored her and continued his tirade at me so that I couldn't get a word in edgewise. Ms. Doolittle did nothing to quiet him or stop him, even though he was being extremely loud and disrespectful. That's when I left, with a departing word to him that he sounded like a moron. Hearing that, he exploded in anger, and charged towards me and the door. I walked back to my classroom and was most of the way there when he exited his classroom and came running down the hall screaming and shouting all kinds of stuff at me. Thinking I was going to be attacked, I

Set Our Children Free

turned around to prepare myself for what was coming, and he got right in my grill only inches from my face. I raised my fists in front of my face in a defensive stance, just like I had been taught in karate many years ago, just in case he decided to throw a punch. The other kids who followed him out the door grabbed him and pulled him away before anything could happen. I went back to my room and wrote up a disciplinary note on him, and assumed he would get suspended and that would be the end of it. Just goes to show you what happens when you assume.

Imagine my surprise when I found out later that I was the one being investigated, and called on the carpet to answer for my actions. I was floored. The principal, Ms. Johnson, had gotten statements from the students who witnessed the event, plus Ms. Doolittle's statement, and I was to write my version also. Why were they investigating ME, I thought. I'm just trying to enforce school rules and this kid goes ballistic, and I'm to blame? The statements turned out to be surprisingly consistent, so there was little dispute about the facts. The only slanted version, in my opinion, was Ms. Doolitte's, who seemed upset that I had disrupted her class. Really? My fault for disrupting her class? It gets better. The investigation was more like an inquisition. There was Ms. Johnson, Mr. Green, and the school resource officer firing away at me. It was like a scene out of a bad detective movie. The only thing missing was the dingy incandescent light with a shade, hanging over my head while the questioners circled me and said, "We're going to make you talk if it takes all night!" I was angry that I had to answer for simply trying to enforce school rules. According to them, my sin was

escalating the conflict by telling Erkel that he was acting like a moron. I could have given them the benefit of the doubt that they were just being politically correct by being concerned that I was hurting the poor boy's **feelings.** After all, if it was just about the "moron" comment, I could have said I was sorry, even though I didn't call him a moron, I just said he was ***acting*** like one. But then they started piling on, and I suspected then that they were out to get me, and get me good. The two other charges were that I had raised my fists in a "fighting" stance, and that I had left students alone in my classroom. This is where the absurd became completely insane. They didn't seem to even want to understand that putting fists in front of your face to defend yourself could possibly be a defensive maneuver. They said they didn't consider it defensive unless the hands were open. I guess they wouldn't have been satisfied unless I had shrieked and held out my hands wide open like they do in B-rated horror movies. Then, I was incredulous when Ms. Johnson asked me why I turned around to face Erkel. At this point I thought that I was in a bad dream, because I couldn't believe I was in the same room with someone so incredibly naïve. When you know that a screaming, angry person is charging you from behind, it is only natural to turn around to meet the threat, or if nothing else satisfy your curiosity as to how badly you were going to be attacked. What was I supposed to do? Run? This was one of several encounters I had with Ms. Johnson that led me to believe that this woman was not dealing in reality, and therefore had no business being a principal of a high school. As for leaving my classroom, I had given the students an assignment and they were busy working on it. I wasn't gone for more than 30 seconds since Erkel's classroom was right next to mine. When Mr. Green told me that I was not even permitted to be in the hallway when I had students in my classroom, I reminded him of his hypocrisy in keeping me out in the hallway for 20 minutes a few weeks earlier when he needed to discuss a student matter with me. Like I said, I suspected that these charges were their effort to get rid of me because I wasn't their kind of teacher. They put a letter in my file as a disciplinary measure. When I asked what punishment Erkel was going to get, they told me he got two days out-of-school suspension. Wow, that should teach him! I'm sure he enjoyed every minute of his time off. I was fuming at this point and told them I wasn't sorry I said he was acting like a moron, because he was.

This incident occurred near the end of the school year, and I should have taken the hint and not returned the following year. I actually considered quitting on the spot, and I knew that would make them happy, but it would be extremely unfair to my students. It was obvious to me that I was not working for rational people. In fact, I was now regretful that I had turned down an offer to teach and coach that year at the school where my wife worked, which was in another county. It seemed like a perfect job – I would be teaching only honors kids in advanced math (calculus) and physics, and would be coaching either varsity baseball, softball, or wrestling. I knew the principal there was a no nonsense type of guy who had a much better grasp on reality when it came to school discipline. Like I said, everything happens for a reason, and I believe I didn't take the job because it was time for my teaching career to come to an end – starting with the above episode. This event was the first in a series of events that drove me out of teaching, which I now look back on as a good thing – for me at least, if not for my students . The following year I got into another confrontation over a CD player with Jackson, a large football player. He came up front and said he was going to kick my a__ for taking it from under his desk, and after a brief wrestling match, managed to get it away from me. I would not allow him to return to my class ever again, but since it was near the end of the school year of his senior year, I did not require him to change his schedule. If I was a vengeful person, I could have, and he wouldn't have graduated with his class. I made some accommodations where I would give assignments for him to do in another room. This is another example of a kid who should have been expelled but got a slap on the wrist instead.

In Ms. Johnson's eyes and Mr. Green's eyes, any conflict between teacher and student was the teacher's fault. After all, teachers were expendable, but we couldn't scar these precious teenagers for life. Two other quick examples come to mind of their obvious coddling of students. The very first day of school Mr. Green stood up at the faculty meeting and the first words out of his mouth were: "You know teachers really do some stupid things. When I hear about a student who loses his temper and cusses out a teacher, I ask myself what that teacher must have done to cause such behavior and set that student off that way." I wish I was joking, but he really did say that. We teachers looked at each other with a look that said, "What color is the sky in this guy's world?" Every teacher in that school knew what I knew – that

Set Our Children Free

teachers just don't get in the face of students and antagonize them into acting badly. Most of the time all that's required to get a bad reaction out of a student is to quote him a school rule that he/she is breaking. They don't need any encouragement to be disrespectful. They had been getting away with it for years, because they knew that there would be no real punishment. I talked with Mr. Green about his comments extensively on other occasions. It wasn't a matter of misunderstanding him. He really did believe that nonsense! He tried to tell me that if I used certain techniques in dealing with misbehavior that I wouldn't have a problem. I fully understood the techniques and had tried them many times. They don't work when you have no-conscience kids who feel entitled to act any way they choose, because they simply don't recognize any teacher as having authority over them. Ironically, this may in fact be partly due to the affective education programs that the schools teach - programs which I discuss in detail in other chapters of this book.

Ms. Johnson, like Mr. Green, was equally clueless about discipline. One of my fellow science teachers, Mr. Lane, once had a girl student that said she felt uncomfortable around him. They had a parent-teacher-principal meeting to discuss it. She was asked repeatedly if Mr. Lane had done anything inappropriate, or had acted in any way to make her feel uncomfortable. She denied that he had done anything to make her feel the way she did. My guess is that she just wanted to change teachers so that she could be in the same class as one of her friends, but that is merely a personal opinion. So they granted her request and she left the room.

After the girl left, Ms. Johnson then told Mr. Lane that they would now have to discuss his punishment. He was flabbergasted! He had done nothing wrong, by the girl's own admission, and yet Ms. Johnson still blamed <u>him</u> for somehow being responsible for the girl's **feelings**. He ended up getting a letter in his file over the incident. He threatened to file a grievance since he was a member of the union, and I hope he did. He also blamed her for taking a coaching job away from him without any good reason. This goes to show to what lengths some administrators will go to indulge students at the expense of teachers. And after treating teachers this way, Ms. Johnson was fond of putting candy in their mail boxes with little notes expressing the school administration's gratitude for everything that the teachers did. I told her before I left the school that the candies in the mail boxes meant nothing if she wouldn't back a good teacher who had been falsely accused by a misguided student or parent.

The fact is that schools are rife with bad behavior. There's the drug abuse, the sexual misconduct, the stealing, the cheating, the fighting, the bullying, and a total lack of self-control – and that's just the faculty! The kids are worse. Long before I ever taught school, there was a candidate running for school board on a platform that the kids were out of control and that there was no discipline. This was in a different county than the one in which I taught. He was a substitute teacher, and I thought at the time that he was just blowing smoke. Once I had taught school however, I knew that this candidate was speaking the truth. Boy did he ever speak the truth! I will make this statement and you may not

Set Our Children Free

believe it when you first read it, but I hope to convince you by the time you finish reading this chapter. Bad behavior and school discipline are not just a problem in our schools; they may be the **number one** problem. Low academic expectations, which I discuss elsewhere in this book, will drain the minds of our students. Bad behavior, unchecked, will drain their soul. The continuous coddling and indulging of repeated bad actors on our campuses not only wastes time and energy from the teachers and the curriculum, but discourages the good students as well. It is very similar to the culture that is created in society when people work hard at building a good life for themselves and their families only to see criminals and politicians (same thing) rob them of their time, their money, their taxes, their dignity, and their peace of mind. They are further discouraged when they see the lack of justice exacted on these same people. The culture in our school system is no different. The good students and the good teachers want badly to have an atmosphere that promotes excellence, instead of one that tolerates foolishness and violence.

I've always thought that there was no reason to have rules if they weren't going to be enforced. The same Mr. Green I mentioned earlier, who felt all student misconduct in the classroom was the teacher's fault, also made it clear he didn't want to be bothered enforcing a dress code. As I mentioned in another chapter, this resulted in an absolute flaunting of the dress code rules. On the other hand, drug and weapons violations were enforced to the letter, or as school officials like to say, zero tolerance. Zero tolerance is practiced by most schools, and every year results in some highly publicized incidents in which students are unnecessarily punished for possessing things like cough drops, a nail file, or anything that remotely could be considered a drug or weapon. Recently I read about a child that was suspended for having a Jolly Rancher![33] They've even gone as far as punishing thought, like when an elementary school kid was disciplined for drawing a picture of a gun,[34] and another suspended for drawing a soldier.[35] I would like to think that these incidents are the exception rather than the rule, as I did not see anything quite that ridiculous at my high school. But because a number of such incidents make the news, I suspect there a great many more that don't, particularly in the primary schools. The incidents of misbehavior I discuss in this chapter, however, are definitely the rule and not the exception.

The main reason so many students misbehave is because they can. Pretty simple. They know there will be no serious consequences and in fact, things like mouthing off to a teacher or some other daring act may actually score them points with their friends for being so cool. I never had much patience for disrespect, and warned the students on the first day of school, both verbally and in my syllabus, that it would not be tolerated. The first year I taught, I didn't know there was a state law giving me the option of removing any student from my classroom permanently for certain behaviors including disrespect. That first year, I had few options. I tried humor, but when that didn't work I would just write them up and send them to the office, until I was told that the school principals didn't like the teachers doing that. That left me fewer options. I tried fear. Sometimes when a student would threaten me, I would give them a cold stare and say, "You know I was in the military. Do you know what they teach you in the military? They teach you how to kill people!" That usually stopped the threat, at least for awhile. Like I said earlier, it is a wonder I didn't get fired that first year. After fear, there were really no other good options, so you called their parents, or at least whoever you got on the phone trying to pass themselves off as a parent. The worse the behavior, the less likely you would get a real parent on the phone when you called, and most of the time it was a grandparent, step-parent, or single parent. Even among the top performing kids, if I did get any misbehavior, it was usually from a spoiled brat girl student raised by a single mom. Rarely did I ever get any misbehavior from two parent families who had the same name as the student. In fact, it was so rare that I don't remember a single instance from any such student in the years that I taught. If that doesn't speak volumes about the cost of divorce, I don't know what does. The biggest kept secret today is that the vast majority of young people who end up being dropouts, teen moms, substance abusers, prisoners, homeless, suicidal, or perpetrators of a serious crime, come from a fatherless home.[36] It doesn't take much of a stretch to extrapolate the fact that these same kids, from the same home environment, are the ones who perform poorly academically — a point I made earlier.

I hadn't been a teacher more than a week or so, when I hear a ruckus behind me as I was passing out some papers. There was Sal and another student whose name I forget, in the middle of a full-fledged fist fight. I immediately dropped what I was doing and waded into the middle, trying to separate them, and yelling at them to stop. Since they were landing some pretty heavy blows, it occurred to me that one of those blows was likely to land on me, since they showed no intention of stopping. Quickly adopting a new strategy that was less likely to result in my face hurting one of their fists, I grabbed Sal from behind and threw him across the room. That at least stopped the fight and I told them to sit down and knock it off. In the background I was hearing whispers from the students. "Whoa, did you see Mr. C throw Sal across the room?" I resumed class immediately like nothing happened. Art, one of the other boys in class, then asked with some measure of surprise, "You mean you're not even going to write them up for fighting?" I said no, that it was more important to get on with class. Like I said, I was one week into the very first semester I ever taught, and knew nothing about what to do in such a situation. Of course I was supposed to write them up and send them to the office, but I didn't know that, and didn't even know what form to use even if I did write them up. Since I realized later that I probably screwed up on the paperwork as well as my violent reaction to the fight, I went to talk about it with Ms. Parsons, one of the few good assistant principals I ever had. She made me feel better when she told me that Sal was a bully and didn't belong in school, and should be kicked out. As it turned out, he didn't even last the remainder of the semester.

This was the first of a number of fights I encountered while teaching. Kids were always fighting about something, and it wasn't restricted to just the boys. I saw some very violent girl fights also. As for breaking up fights, I found out that there were two different approaches teachers would use. One would be the passive approach, which was to do nothing and call the front office for help, hoping someone would get there before the kids killed each other. The theory here is that any contact with the students could get you hurt or get them hurt, and get you sued. The other approach was the one I used, which was to break up the fight and worry about the consequences later. The theory here is that if you stood by and didn't break up the fight when you had an opportunity to do so, and some student got hurt as a result, you could get sued. There you have it. Pick your poison. You could get sued either way, but I always felt that I just couldn't just stand by and watch someone get hurt, so I erred on the side of breaking up the fight. One teacher told me that when he had to break up a girl fight, he would pull them apart by their hair, so as not to leave a mark, and thus not be accused of using too much force. One time while I was on lunch duty, I heard some jawing, so I looked up and saw one student marching in a menacing manner towards another who was backing away. There was a big crowd forming, and it was clear that the aggressive student's intentions had nothing to do with introducing himself and sharing yearbook photos. I ran toward the confrontation and the closer I got the bigger the aggressor looked. When I finally got between them and put my hand out and yelled "stop," I found myself looking up. Way up. This was the biggest high school kid I had ever seen. He must have been at least 6'8" tall and over 300 pounds. My eyes were about the same level as his solar plexus, which is where I aimed my extended arm and braced for the collision, since he was still coming forward and talking trash. He bumped into my arm and stopped cold. At times like this, I was grateful for having been a pretty good athlete myself and learning a little about leverage. I yelled at both kids to go their separate ways, which they did. Then I heard the whispers again from the crowd. "Whoa did you see Mr. C stop Brandon? He was charging, and boom. It was like he hit a brick wall!" They were still talking and asking questions about it a couple of days later. For my sake, I'm just glad Brandon did the right thing and stopped, because I certainly wasn't looking forward to a wrestling match with him, and I can understand why the other kid was backing away.

One time I was talking with another teacher outside the science building about his approach to breaking up fights, which was just the opposite of mine. He believed in the passive approach. As we were standing there talking about that very subject, a fight broke out just a few feet from us. I jumped into the middle of it and pulled the kids away from one another while this teacher just stood there. I don't blame the other teacher, as he genuinely felt it was the right thing to do. I just disagree. If it was my son or daughter involved in a fight, I would like to know that responsible school officials would do everything they could to protect them. The most violent fight I saw was in the cafeteria one year when two kids who were brothers jumped another kid for some grudge that had been brewing. I looked up from reading my newspaper to see students scattering as the two boys were swinging at another who was doing his best to try to protect himself. Before I could get to them, they had punched him several times, and he fell backward and hit his head on the hard tile floor with a sickening thud. I was surprised it didn't knock him out, although I'm sure he had a serious concussion. Coach Larson got there a split second before me, and we both wrestled the perpetrator to the ground. The coach was in fact a wrestler in high school, so he not only got him down, but had him locked in a pretty secure hold. The other brother had given up the fight and didn't need restrained. It seemed like only seconds later that Officer Prince, the school resource officer, came and put handcuffs on the boy. This was a welcome sight to me. For several years, this case would have been dealt with by suspending both parties involved in the fight – aggressor and victim. In fact, that is the way most schools handle fights. Most schools don't recognize a student's right of self-defense, and tell the kids that when confronted by bullying, they should just submit. This is one of the most foolhardy policies any school can adopt. It leads to more bullying, and in some cases, school shootings when the victim gets tired of it. Schools really favor appeasement over confrontation – a kind of feminization policy that is incredibly naïve. It springs from the left-liberal notion that if we're just nice to people (and other nations), they will just leave us alone. Also, some school officials are just too lazy to investigate and find out who really was at fault, so it's easier just to punish both. Many schools' response to bullying has been to introduce an anti-bullying curriculum – a convenient excuse to promote sensitivity towards all the politically correct protected species on campus. Even the feds want to get into the act with such legislation, and to further protect against cyber-bullying. These are all wrongheaded approaches that choose to restrict the rights of all students rather than deal with the root of the problem – the bully. My son was suspended from school one day because he shoved back when another boy shoved him. When I asked the principal why she suspended him, she said that he could have gotten hurt fighting back, and that he needed to be taught that when he grows up it could cost him his life if someone picks on him and he doesn't submit. Again, I wish I was kidding but she really did say and believe that. These are the kinds of people running our schools. That's why I was glad to see the handcuffs from our new school resource officer, who did not put up with such nonsense. I asked him later what would happen to the boys and he told me that charges would be filed against the aggressors, and nothing would happen to the victim. Praise the Lord for small victories. As I said earlier, the stifled sense of justice that prevails at most schools is one of the things that destroys kids' spirits, and leads to an undercurrent of seething anger that sometimes manifests itself in reactive violence. The solution to bullying is really not all that complicated, and I address it in the last chapter.

It's funny how girls and boys have different reasons for fighting. Boys are a little more complicated in their reasons, but not much. Many of them have anger issues because of their home life, and just want to strike out at someone. Sometimes the fights involve drugs or girlfriends. Most of the time, however, it's a plain and simple testosterone-filled, mano a mano, pride-induced desire to be a man and not take what they deem is a slight or disrespect from another male. Some guys never grow out of this stage, and even as adults, are ready to "go" at the slightest perceived insult or even a wrong look in their direction. I think movies play some role in this, as many of the action heroes are guys that can whip three other men with their hands tied behind their back, and still get the girl in the end. Too many guys are still trying to be that macho man that never loses a fight, although most guys eventually get to be comfortable in the assurance that they don't have to be that unrealistic icon to be a man. Girls are somewhat simpler when it comes to fighting – they fight over boys. In fact, I've never seen a girl fight that wasn't about a guy. If they only knew how foolish they look when they are pulling each other's hair and trying to gouge out each other's eyes. Get a clue girls. The girl you are fighting is not going to be driven off or intimidated by you if your guy wants her. And your guy is not automatically going to go home with the winner of the fight like a buck following the winning doe. If he's attracted to the other girl, there isn't a blessed thing you can do about it. Besides, I never knew a guy who said, "Wow, man she's hot. Look at her throw that right hook! That's my kind of girl."

Speaking of fighting, one time I almost got arrested by the same Officer Prince, for somewhat losing my temper and saying the wrong thing to a student. I was talking with Sabrina, one of the girls in the front row, and apparently I had said something to Randy just moments earlier that had upset him, unintentionally on my part. I then heard him threaten me loud enough for the entire class to hear. This ruined my good mood and I said something to him about knowing when and where I got off work if he wanted to do something about it. This got me into a conference with the principal and Deputy Prince, our school resource officer, during which both Randy and I rightly apologized to each other. The Deputy felt the incident didn't sufficiently rise to the level of assault, and he was right, but it was still stupid on my part. Several of the students in the class told me later that Randy was steamed because Sabrina was an ex-girlfriend of his, which I didn't know, and he was jealous that she was paying attention to me.

We never had any school shootings while I was there, but we did have two teachers from our school shot and killed by a student near the school property a couple of years earlier. And we had the usual threats of shootings. One in particular I remember was on the last day of school. We had received a phone threat that some student was on his way to school with a gun, and the teachers were assigned duty stations around campus to look out for such an individual. Now if this strikes you as pretty silly, then you are possessed with a great deal of common sense – something very lacking among our school administrators. What were unarmed teachers supposed to do if they ran into an armed student? Talk him out of it? Run and yell for help? Take a bullet for the team? Not me. If I'm prohibited from carrying a gun at school, even though the state trusts me to carry one everywhere else, then don't expect me to be one of those sacrificial heroes who died while trying to talk sense to an angry, armed student, leaving my wife and family with nothing but memories. So when we received the threat on the last day of school, did I report to my duty station? No way. I was standing right next to our school resource officer, talking with him the entire time, since he was permitted to carry a gun and several magazines of ammunition. The threat never materialized, but this was just one of many times that I marveled at the stupidity of how these situations are handled. During school hours, anytime there was such a threat, we went into lock-down mode. This is pretty standard nationwide, and equally stupid nationwide. We are required to lock the classroom doors and stay there like sitting ducks. Since the doors are partially glass, anyone with a gun can enter any classroom he wants. I've told my students more than once that if a gunman shows up in the hall outside our classroom door, don't block my way to the back window, because I'll be the first one out. Let me carry a gun, however, and I'll stay and fight for their lives. All schools are inherently unsafe for the simple reason that everyone knows that they are gun-free zones.[37] Change that equation, and you will make it much riskier for the shooter, and safer for the students. Teachers with guns? Oh, how horrible! No, how sensible! It has successfully deterred gun violence in Israel and Thailand, where they have to deal with the threat of terrorist attacks as well as student violence.[38] This politically correct attitude even affects the military. During the Ft. Hood shootings in which 12 people were killed and 31 wounded on Nov. 5, 2009, not one of the victims was carrying a weapon with which they could defend themselves. Why? Because regulations prohibited soldiers from carrying personal weapons on base, thanks to one of the very first acts signed into law by Bill Clinton![39] Say what? A soldier can't carry a gun on base? What foolishness! The gunman obviously didn't obey those regulations! Does anyone think he would have stopped and thought, "I would really like to kill all of these soldiers, but I can't because I'm not allowed to bring my guns onto the base?" Gun laws only restrict law abiding citizens; they do nothing to inhibit those intent on criminal activity. If just one of those victims had been carrying a personal weapon, they could have saved many lives. They're in the Army for crying out loud! Why can't they be trusted with a gun on base? Does any sane person seriously think that a shooter contemplating mayhem will change his mind when he finds out he's not allowed to have a gun on the premises where he intended to slaughter people? I don't want to belabor the point, but the highest crime rates are found in the cities and states with the strictest gun control laws, contrary to the news media version that more guns equal more crime.[40]

Bomb threats were another irritating trend we had to put up with. I say irritating because when was the last time you heard of a school being bombed? Usually there is a 99.99% chance that it's only a threat, yet every time there is a threat, we have to take it seriously. It didn't take long after I had started teaching my first year before I got the first bomb threat. Usually they were called in by phone from a student who was absent that day, and some students are dumb enough to call from their home phone, which was easily traced. The first one I got, however, was from a note that one of my students "discovered" on the floor of the classroom. It was a bomb threat all right, but written such that I knew immediately that it was a hoax. Unfortunately, I also knew that I would have to report it, and lose several hours of valuable class time during which the students had to remain outside the buildings while a bomb squad searched. Of course, that was the goal of the student who wrote the threat – a nice break from class and/or a scary test that day. I thought it was a good idea that the students were required to take their back packs with them, which cut down on the search time. I got two more threats that year, but never again got another. If I'm not mistaken, it was the SIXTH PERIOD FROM HELL where all the threats occurred.

Of course fighting, bullying, and bomb threats aren't the only diseases that drain the life and vitality out of our schools. Drugs are bought, sold, and used daily on our campuses. I don't think this is any mystery to anyone, but my point is that there is an undercurrent of lawlessness that school administrators seem powerless to tackle. This is despite the presence of a school resource officer on most campuses, and a no tolerance policy. Kids will not turn in other kids in for drug use or their lives will be threatened in many cases. And teachers are simply not devious enough to catch them most of the time. The safest place to have a smoke or get high is, of course, the restrooms. When I was on a school advisory committee at another school in another county from the one in which I taught, there were constant complaints by students that they couldn't use the restrooms because of the ever present smoke. When I actually taught school many years later, things hadn't changed much. It was too hard to catch the students because teachers generally couldn't spend the entire recess or lunch period in the restrooms. It got to be such a problem at my school that our administration decided to lock all of the restroom doors and prohibit the students from using the restrooms between classes. Therefore, the only time kids could go to the restroom was <u>during class</u>! Think about that. When students should be able to use the restrooms, they couldn't because of the rampant drug and cigarette use, not to mention sexual escapades. Who did this rule punish? It was the good kids who were made to suffer for the deeds of the bad, because they couldn't use the restrooms for the purpose for which they were intended, and at the most convenient time. So that left only valuable classroom time for most kids to use the restroom. For those classrooms that didn't have a restroom in the classroom itself, a teacher had to have sign-out sheets and restroom keys to use one of the other restrooms in the building. And for obvious reasons we could only let one student at a time use the restroom. That meant that some students had to wait their turn and miss a portion of the lesson just to go to the restroom. This is another example of how bad behavior robs good kids of an opportunity to learn. One year my classroom was directly across the hall from the restroom. It was not unusual to smell cigarette or marijuana smoke coming from that direction. One time when the smell was particularly bad, I noted the three students who had just left the restroom. I then called Coach Larson to go with me to their classroom and call them out of class to search them. One of the boys I knew was a known drug user. I thought we would find something but we had them empty their pockets, wallets, etc. and found nothing. I'm sure they were smoking both tobacco and weed, but they tended to be ingenious about hiding it. Most people don't realize that a school teacher or administrator has more power to search than a police officer. The Constitutional restraints that restrict when, how, and for what reason a policeman may search an individual have been somewhat diluted by the courts for the special circumstances found in a school situation.

Set Our Children Free

I briefly mentioned an incident in another chapter, where a boy had pulled a girl's skirt down to her ankles, exposing her since she was wearing no slip and only thong underwear. The girl, named Adrian, was somewhat traumatized and made it clear to me that she wanted it reported. In other words, this was no friendly prank. It may have even constituted sexual battery. I reported the incident to Mr. Green, the assistant principal, both by phone and e-mail, along with the perpetrator's name. I assumed the matter would be taken care of with the boy being suspended, and possibly charged with a crime. I ran into Adrian a few days later and asked her if she had been contacted by the principal about the incident. To my surprise, she hadn't, so I wrote a disciplinary note on the boy, and once again contacted Mr. Green, thinking he may have been busy and forgotten the incident. Adrian made it clear that she still wanted to pursue the matter. A few weeks later I ran into her again, and again she told me that she had not been contacted and the boy had not been disciplined. By this time, however, she felt too much time had passed, and she just wanted to drop the matter. She had lost faith that the school officials who were there to protect her were up to the task. This both angered and saddened me. I felt like I had failed even though I followed proper procedure. If I knew then what I know now, I would have reported it directly to the school resource officer. Later that year I reported the matter to the head principal, Ms. Johnson, and still got no response. Finally, when I decided to leave that school forever because of the total absence of any student discipline, I wrote a letter describing the incident and copied each member of the school board and the Superintendent of

Schools. Once again, I was totally ignored. Not even one official was curious enough to call me and ask me about the incident. It's possible that one of them may have called the principal first, and got the standard garbage about me being a disgruntled former employee, etc., but no one ever asked me for my side of the story. How could two different principals, a Superintendent of Schools, and an entire five member school board all be so derelict in the duty that they were sworn to uphold? In my opinion, there should have been an investigation, and both principals fired. Yet the extent of their investigation didn't even go as far as picking up the phone and giving me a call.

I was a little naïve about how many thieves existed among the student body my first year. I already knew students stole from each other, as my own kids would regularly "lose" valuable things when they attended school. I didn't know, however, how much stuff is stolen yearly from the school itself – stuff that the taxpayers would have to replace. I was teaching a science class one day early in the semester, and this particular day was a lab day. There are lots of interesting gadgets in a physical science lab that students love to play with – like strobe lights, electric motors, magnets, generators, etc. At the end of class, students were responsible for cleaning up and returning the lab equipment to its proper place on the shelves in the back room. This was the first year I had taught science and I wasn't yet aware of just what the kids were capable of. When the bell rang to end the class, two good responsible students from that class came up to me and told me that three of their classmates had made off with several of my $250 strobe lights in their backpacks. The kids explained that the only reason they were telling me, was that they elected to take my class because they liked science labs, and that they were afraid that when I discovered that the strobe lights were missing, that I wouldn't have any more labs. Their fears were well founded of course, which is another example of how bad student behavior robs good kids of an education. I got the names of the kids who stole the lights and told the good kids I wouldn't use their names. It wasn't until later that I got to talk with the school resource officer, but I was fuming when I did. I have always hated thievery, and I wasn't about to put up with it. We called the bad guys into the office the next day – one at a time, on purpose. Divide and conquer was the plan. We told them each that we knew they stole the strobe lights. We said that if we got them back, no problem. But if we didn't we would have them prosecuted and their parents would have to pay the cost. We also told them that if they admitted the obvious, we wouldn't tell the other guys in their posse that they had confessed. After a few rounds of questioning, we got the truth and the strobe lights back, except for one that they threw out the car window on the way home. The parents had to pay for that one, and the kids did get written up, but at least they didn't get prosecuted. After that incident, the students' freedom to retrieve and return lab equipment was severely curtailed. I always locked the back room even when I was in the classroom, and I locked all of the cabinets that faced outward towards the students. Then, on lab days I would set up the equipment myself even if there were a number of stations to do. This cost a great deal of extra time, but I couldn't justify getting valuable equipment stolen, because there may not have been enough money in the budget to replace it the following year. Even after these precautions, I still had things stolen, as some of the gadgets were pocket sized. Because of block scheduling, some of the lab equipment had to be left out on the lab tables for two days so several different classes could use them. In the meantime, one or two non-lab classes would meet in the same room. So there was ample opportunity when my head was turned for students to steal lab equipment lying around on the tables. The alternative was to put everything away after each class and inventory it before the students left the classroom, which was completely impractical.

I've found that kids are so afraid to rat on one another, that you pretty much have to promise them anonymity. When students do have the courage to speak out against improper behavior, I don't feel that they should be put into a position where they are targets for retaliation. I used this principle throughout my teaching career, and I got a lot more cooperation from students willing to report things. One time I had a class with a few ornery kids that showed zero respect for the teacher. I was writing on the board with my back turned to the class when someone threw a spitball at me. It was my first year teaching so I didn't quite know how to handle a situation when I couldn't identify the perp. I turned and stared but then continued the lesson. In another thirty seconds, here came spitball #2. Another stare, and then #3, #4, and #5 in short order. I knew I had to do something quick because it was getting contagious and then it would be out of control. I could have just stopped class and not taught any more that day, but that would have punished the kids who really wanted to learn, and then there would be no guarantee that it wouldn't happen the following class. Situations like this can either make or break a teacher so I had to think quickly. Fortunately the answer came to me on the spot. I told the class that if there was one more spitball, that I would stand outside the classroom and call every student out one by one and ask them secretly who did it. I told them I knew there were kids in the class who would finger the miscreant, because I wouldn't identify the one(s) who told me. And when I found out who the bad actor was, I would make sure that they were suspended for a long time. This wasn't a court of law, and they didn't have the right to face their accusers. They knew I was right, and that ended the spitball attacks. When I told the story to several of my fellow teachers, they told me that they never turn their backs on the students. I then asked how they wrote on the board and they showed me in their best sideways contortionist stance how they did it. One teacher told me she never used the board; she only used handouts. Since you can't use handouts for everything, and I was never very good at writing on the board without looking at what I was writing, I just continued using the board the way I always did and never had another problem.

Another theft incident I clearly recall was when I was coaching basketball. The school had just bought about a dozen to fifteen new basketballs for the beginning of Fall practice. At the end of practice each day the balls were stored in the coach's office locked behind a steel door. One day we came in and they were all gone with the door still locked, and the thief was never found. As happens so often in cases like these, the balls were never replaced, so we had to use the old substandard basketballs the remainder of the season. Once again the good students suffered for the deeds of the bad. The most frustrating thing to me was that there were only a few people who even had the keys to that office, and no investigation was ever conducted. I had to be extremely careful about my personal effects also. I always locked my wallet and briefcase in a file cabinet in my room, and then locked my room when I wasn't there.

Cursing was also against school rules, but it was another one of those rules that was never enforced. I simply got tired of not being able to walk across campus without hearing foul language continuously. Kids knew that you weren't going to write them up, because outside of the classroom, teachers didn't carry disciplinary forms with them. At least walking across campus I only had to hear it once from each group of kids I passed. When I was on lunch duty, however, I couldn't escape it because I had to stay in a specific area of the campus where the kids would mingle. In those instances, I made up my mind I wouldn't tolerate it, so I told the kids ahead of time that if I heard certain words, they would have to leave the area. I wasn't about to carry disciplinary forms around with me, but at least I could banish them from my hearing range, and they wouldn't be able to spend the rest of the lunch period with their friends. They hated this of course. They would argue that everyone curses, and that it wasn't fair to single them out. I would point out that they were in school to learn that society considers that kind of language unacceptable, so they might as well learn that now. I said that they wouldn't use that kind of language on a job

Set Our Children Free

interview, so they couldn't use it here. The Federal Communications Commission doesn't even allow it on the airwaves. They continued to curse, and I continued to banish them from their favorite lunch time hangout.

Lying from students was another thing that I found difficult to tolerate. I told them from day one, that if they misbehaved, and I caught them, that they should just apologize and we'd move on. I had no interest in embarrassing a student. If they were talking while I was giving instruction, for instance, I would stop talking and look at them. If they stopped, that would be the end of it. If it happened again, then I would say something or move them to another seat. If they argued with me, however, then we entered a new arena. When I was in school, and the teacher caught me doing something wrong, I was embarrassed and humbled. Not so today. Most students' first response is a lie such as, "I wasn't talking." I cannot begin to tell you how many times this happened. If teachers could fine students a dollar for every time a student was caught lying, they would soon be highly paid professionals. Sometimes I would get sarcastic when I received a response such as this. I would say, "Do I look like I'm deaf and blind?" Believe it or not, that wasn't enough to convince most of them to just admit the obvious and shut up. Now they had to defend their honor in front of their fellow classmates, so they would start to get ugly, which precipitated them getting booted from class with a write up. Sometimes I would give them a few moments to think of the consequences of lying to me by saying something like, "Think about what you're denying. I am not deaf or blind. If you just say you're sorry and not do it again, we can continue class like nothing happened. But if you insist on lying about it, you'll be gone. Now after careful consideration, what was your answer again?" A few students would relent and say they were sorry and that was the end of it. But believe it or not, most students persisted in clinging to the lie. They would look me right in the eye and tell me they weren't talking, weren't cheating on a test, etc. and expect me to deny what my own eyes, ears, and logic were telling me. I've often wondered why, until I realized that they operate on a different reality than honest people. Honest people realize that there are fixed standards that we all must adhere to in a free society – that principles of truth, justice, trust, and respect are absolutely essential in both interpersonal relationships as well as business and commerce. To many teenagers, the line between the truth and a lie is nonexistent. Reality is what they say it is because it is all about what is right for them. If lying gets them what they want, then it's perfectly OK. This is what happens when we have no fixed standards of right and wrong, no moral compass, only moral relativism – the only kind of philosophy a kid can logically adopt, by the way, if he's taught evolution or the principles of Outcome Based Education . Lying, cheating, and stealing are acceptable if an individual can get away with it, and most of the time they do, which encourages more of the same. And they didn't just lie about misbehavior; they constantly lied to their parents about grades. Good grades mean privileges, bad ones mean restrictions. If they had the kind of parents that linked grades to privileges, and most did, then bad grades meant staying at home in many cases. Girls especially freaked out when they found out what they were going to get on their quarterly grade because they were about to experience the end of their social life as they knew it. Most of the girls who felt like this were unsurprisingly the type who really needed to be home on the weekends anyway. The only way out of this mess, in the mind of the student, was to blame the bad grade on the teacher, thereby transferring responsibility to someone other than themselves. If successful with this strategy, freedom restored! So I would get a lot of calls right after grades went home from parents who were convinced that their kid really had turned those papers in, or that the last test was so unfair, or that their kid really was in school all those days I had marked them absent.

If I was intolerant of lying, disrespect really sent me off the cliff. After my first year, when I put up with enough disrespect to last a lifetime, I was determined that it wouldn't happen again. I had sacrificed too much as a teacher to be verbally abused by a teenager who knew nothing. So starting with my second year of teaching, I used a new approach. The first day of class the students received a syllabus which detailed, along with the academic requirements for that particular course, my class rules. This syllabus had to be taken home, read, and signed by their parents, so that when I enforced the rules there would be no argument that they didn't have fair warning. One of the rules was that I would respect every student, but I would expect the same from them. There was to be no arguing about the rules after the fact, only compliance. And if their behavior reached a certain level of disrespect – and I usually drew the line at being cussed out – they would be removed from my class forever. Under state law and existing school board policy, we were allowed to do this, but school principals hated enforcing this loophole like the plague. They tried to get me to change my mind, or tell me I couldn't do it. One time I had to write them a letter detailing chapter, line, and verse of the applicable statutes and rules, and didn't get questioned after that. What really upset them is that I was the only teacher at the school who had the courage to use this lifeline. Other teachers were amazed that I got away with it, but since they were career educators and their careers depended upon their principal's favorable performance evaluation, they were too intimidated to do it. I could do it only because I didn't have to worry about losing my job. I felt sorry for some of the other teachers who, like me, didn't tolerate disrespect well, but had to put up with so much from the students. I could see how the stress wore on them.

My view was that if we don't allow paddling anymore, and you can't deduct points from a student's grade for misbehavior, what option is left for the teacher to control behavior? Parents aren't usually helpful, because a misbehaving student usually meant an uninvolved parent. And school administrators didn't help by giving a kid one or two days of in-school suspension (ISS), which meant they got to hang around with their friends in another classroom and do nothing. I found out later that in such instances, we were actually supposed to send the suspended student their assignments while they were in ISS. I refused to do it. Since this was more work for me, I wasn't going to punish myself for something they did. The kids certainly didn't consider ISS punishment, and certainly not a deterrent to future misbehavior. Many school districts still employ paddling, and some are considering reinstating it because of the discipline problems.[41] I'm not sure if paddling is the answer, although I'm sure it wouldn't hurt if it was fully supported by state law, since the last thing our schools need is a flood of lawsuits. I know it was a deterrent to me and my classmates when I was in school, as one paddling was all I needed to straighten me up. The male teachers back then did not give out "love taps," and there was a lot of discussion among us as to which teacher gave the most painful swats. In fact, I don't remember a male middle school teacher who <u>didn't</u> paddle, but I don't remember any paddling in high school. The most important thing to note these days, however, is how much unproductive time is spent during the school day dealing with student discipline problems. This robs good kids from getting the education they deserve. In my mind, any student who curses at a teacher should be given 10 days out-of-school suspension (OSS) on the first offense, and expelled for the second offense. These kinds of kids need to be in an alternative school. Put the fear of God back into them, or they simply won't respect the teacher's authority. The same 20% of kids get disciplined over and over, consuming 80% of the school's time. A student could get disciplined 100 times during the school year, and on the 100th offense, he would get the same punishment as the first offense. This was one of the biggest problems teachers had with the entire disciplinary system – there was no allowance for <u>cumulative</u> offenses. Punishment needs to be cumulative, like points on your driver's license. If you accumulate too many points, even the most irresponsible drivers will take notice and ask themselves whether they really want to lose their license. If we do this for driving privileges and even in our criminal justice system, why don't we do it in schools? Why don't we consider education a privilege, instead of a right? Lawmakers need to re-think the notion that all kids must be kept in public schools until they arrive at some magical age, no matter what they've done. Last time I looked, education wasn't a Constitutional right. And federal law will not allow suspension of a special education student for more than 10 days under the Americans with Disabilities Act (ADA), regardless of whether the student is a danger to others or not.

When I would have to kick a kid out permanently, the other kids in the class would always insist he or she was coming back. Since I was the only teacher who did this, they were so sure I wouldn't get away with it. I assured them I would, and they were shocked when they found out their fellow classmate was actually gone for good. Sometimes I would send the misbehaving student to the office with a note that I would not accept them back, and for the counselors to start re-arranging the student's schedule. The principals were always appalled at this because they were so concerned about the student's "due process." They would insist that there was a process that had to be followed in order to remove the student, which was true, but I made it clear that until that process had run its course, which could take weeks, the kid would never darken the door of my classroom again. Sometimes they would send the student right back to my class, and I would send them right back to the office. This would happen two or three times in a row in some cases, but I was extremely stubborn about this. It was funny in a way.

Set Our Children Free

The ostracized student would come back to class all cocky and arrogant and tell me that the principal said to let them in. I was equally adamant that they weren't getting in. If the other kids in my class saw that I didn't mean what I said, and that I tolerated disrespect, then I would lose the rest of the class, and no one would listen to me. But it was interesting to see the kids' reaction to all of this. They were so unused to being treated this way, even for the worst offenses. In fact, the students were so indulged and pampered at our school, that they thought nothing of getting up right in the middle of class and going to the office without asking permission if they disagreed with anything I said or did as a teacher. They knew they would find a sympathetic ear in the principal's office. Since technically they had to have my permission to leave the classroom, I would usually write them up for this, which usually got ignored. I know the principals didn't mind the kids coming into their office and spilling their guts about the horrible treatment they were getting at the hands of Mr. C, but what if they weren't going to the office? What if they were secretly meeting someone on campus instead, and just pretending to be mad at me to get out of class? And yet I cannot remember one student ever being punished for leaving class without my permission. That is how little respect our principals had for the teachers.

So what is the main reason teachers receive so little respect from most students? That is an easy answer for anyone who has taught school – teachers have no real authority. They have been neutered and stripped of their authority in the only two areas that count: (1) control over their classroom, and (2) the ability to fairly assess (grade) their students in a meaningful way. Regarding control of the classroom, teachers should not have to earn respect from their students; they should be accorded respect by definition. Call me old school, but the only reason any teacher got our respect when we attended high school a generation ago was because they had authority, not because they were nice to us. If we messed with them, talked disrespectfully, or willfully broke school rules, we would have been paddled, suspended, or expelled on the first offense. We feared the teachers because they had the power to discipline us, and there was no running to the principal to save us. In fact, if the matter went as far as the principal, then we were really in trouble. Today's kids run to the principal to save them <u>from</u> the teacher. The other authority the teacher no longer has is the freedom to assess their students in a meaningful way. By that I mean that the unspoken message in every school is that if you grade too hard, and most of your kids are flunking, then the teacher is at fault. Kids inherently understand this, so give very little effort. Recall the true story at the beginning of Chapter 2. They know that most teachers will compromise and adjust so that most of the class has an acceptable grade. Teachers are, in fact, encouraged to do this in a number of ways – grade tests on the curve, count tests as only a small percentage of a student's grade, give "group participation" grades, or give a lot of credit to things that can be copied like homework, reports, and projects. In all of these cases, very little learning takes place. This is why our kids score so poorly on standardized tests, and why there is so much stress when the kids take them. They're not used to having tests actually count for anything. I actually read a pitiful article the other day written by a local school board member lamenting our emphasis on standardized tests, and our setting aside too little time in our schools for teaching "creativity and innovation" (read more non-academic nonsense). And, pray tell, how would we measure, and thus grade their creativity and innovation? Believe me, a teacher who drills their students on the core principles of the course material, requires them to have to study in order to learn those principles, gives tough but fair tests to assure that the material was learned, and then uses the results of those tests as the only basis to assess their students' grades, will likely be looking for another job in short order. I know because that's how I taught school. I know other teachers who were reassigned, required to loosen their standards, or fired for the same reason. Unless we re-empower our teachers with the authority to both control their classrooms and assess their students fairly, we have no hope of improving our schools.

Another school rule which I also dealt with in my syllabus, was that food and drink were prohibited in the classroom. The reason for the rule was obvious. You couldn't have serious instruction going on while kids were woofing down potato chips and sodas, let alone a written assignment or quiz. On top of that, where do you think the potato chip bags, soda bottles, and plastic wrappers got left once the bell rang? If you answered the garbage can, you've never raised teenagers. Nope, they were simply too lazy to walk to the garbage can, so all of it got left right on the floor or under the seat. By the end of the day, I felt like I was sitting in the middle of a dumpster. I am not kidding; there was that much accumulation of garbage in just one day. But, you say, there was a school rule against bringing it into the classroom right? And you re-emphasized it your syllabus? And with block scheduling, the kids had 20 minutes between classes so they had plenty of time to junk up between classes, right? So why wouldn't the kids follow the rules? Seems like a no brainer. This would be of course if you assumed that those who invented the food/drink rule would have the courage to enforce it. I say courage because I've found that there are two sacred cows when it comes to things kids hold dear to themselves, and they won't stand for anyone messing with them – their food and their music. Take their boyfriend or girlfriend, but don't mess with their sodas, snacks, or songs. The angriest I've ever seen students get is when someone takes their food or drink for which they just deposited their hard begged-for cash into a vending machine, or when someone takes their portable CD player/i-pod. Why have rules you're not going to enforce? You figure it out. I never did. All I can tell you is that it caused me plenty of drama. It would start with warnings. I would observe a student come into class with food or a beverage, and remind them of the rule. Oops! They would retreat to the hallway and put it in their backpack or purse. Now realize that this didn't mean that they had abandoned their plans to eat or drink in your class that period. No, no, no! They didn't just purchase those precious commodities to let them sit out of sight in the dark for the next hour and a half. It only meant that now they had to sneak the stuff into their mouth without me seeing it. Once that happened then they could leave the empty bag/bottle where they always did – on the floor or under the seat. About halfway through the class, most kids would relax and bring their food or drink out, confident that

Set Our Children Free

they had gotten away with it. As I walked down the aisles, I would simply pick it up and throw it away, even if it was a full bottle of soda. Of course, they would come unglued when I did this, but I didn't care. They were repeatedly warned and had no excuse. The one food item I hated the worst was sunflower seeds, and unfortunately these were the hardest things to spot while they were being eaten. The kids who ate these were the worst offenders and I'm convinced that they sold their soul to the devil, since I never once was able to catch the perpetrators in the act. The bell would ring to end the class, and then and only then, would I see them – the seed remains all over the floor or under the seat in a disgusting mess. When I would ask the kids why they were such pigs, strewing garbage in their wake everywhere they go, they answered honestly for once – "They hire janitors to clean up the room," neglecting the fact that it would make the clean-up person's job so much easier if they simply discarded their stuff in the trash where it belonged. So monitoring the students as they came in didn't work. I also tried another strategy, where I would just assume that they would bring the stuff into the classroom, and get them to clean it up five minutes before the bell. This was also unsuccessful, because a lot of times there is a lot of activity near the end of class due to labs, tests, and in-class assignments. Stopping the class five minutes early just for clean-up was not always possible. When I did, and I found trash under a student's desk, guess what answer I <u>always</u> got? Correct. "I didn't put it there," or "That was there when I got here." They were then informed that the previous class had already cleaned up before they came in, and guess what their response was to this news? Before you guess the response, read the earlier section I wrote on lying. Yes, they <u>swore</u> it wasn't their stuff. Finally, I just made a rule that whatever seat you were assigned, you owned. If you didn't want to be held responsible for cleaning up trash someone had left from the previous class, then you were responsible for inspecting the seat or area you sit in before you sit down. If there was trash from a previous class, they were to let me know immediately at the beginning of class, and they wouldn't be held responsible for it. This still didn't keep trash out of my room, but at least with this latest method, I could assign responsibility. And the reason it didn't completely work was because most students refuse to accept any responsibility.

For most kids, even for those with bad home lives, the world exists to fulfill their needs. Many of these kids know that it's all about them. It's their world and we are just living in it. From the yearbook ads they buy, to the Facebook, My Space, and U-Tube bios they produce, to the clothes, cars, and i-phones they carry, they feel entitled. I can remember one of my worst students coming to class one day with a new $250 PSP - portable play station. I asked myself what parent would buy an expensive gift like that for a kid that spent absolutely no time on his school work. Where were their values? I can count on one hand the number of high school kids I've met that I would actually classify as humble, and I can't remember any that were camera shy. This indulgence develops an attitude in them that perverts all sense of fairness, compassion, and justice. If I were asked what phrase I heard more often than any other in my years of teaching, it would be, "That's not fair," followed closely by

"That's just wrong" and "That's messed up." Of course, they had no sense of fairness, let alone right and wrong, because their frame of reference was their all-important subjective feelings, not any fixed standard like a school rule. Every time I heard one of those two phrases, it was in response to me doing something like enforcing a school rule or doing something most people would describe as common sense. Yet in their minds, it was unfair and wrong simply because they didn't get what they wanted. I sometimes would tell them that life isn't fair, get used to it. But they've been programmed since primary school to believe that all outcomes in life should be fair, regardless of whether or not it was earned or whether their behavior merited it. Any doubt as to how these students will vote when they turn 18? Again, the good kids from good homes didn't think this way, but they were a small minority of the student population.

In addition to twisted individual thinking, there was twisted group mentality. Kids, like most people will do things in a group (or mob) setting that even their logic would dictate was a bad decision, not to mention that there's more courage in a group setting. Our campus had the usual factions: preppies, blacks, Hispanics, rednecks, jocks, Goths, God squaders, etc., and occasionally there would be group confrontations. School administrators would always overreact of course, because the number one job in their mind was to keep peace on campus. Part of this overreaction was to remove anything whatsoever that could possibly offend one of the other groups, such as displaying a rebel flag on a T-shirt or ball cap, for instance. I've heard of other school districts that banned anything that had a religious symbol or phrase. Although courts have ruled against these kinds of restrictions, that doesn't stop school administrators from trying. Remember, these are the same people that preach tolerance and respect for diversity, but feel no compunction in abridging a student's free speech rights. If the Holy Grail is tolerance and diversity, then why not require students to respect <u>all others'</u> viewpoints, whether they agree with them or not? The dirty little secret is that the only viewpoints school administrators are intolerant of are conservative viewpoints. Students who object to these viewpoints should be told to be tolerant, just like most students are told when they object to being taught sexual diversity or multiculturalism. Instead, hypocritical school administrators only believe in tolerance if they agree with the message to be tolerated. To them, offense is defined from the viewpoint of the listener or viewer. When that is the case, then a person's freedom of speech belongs not to them, but to whoever hears the speech. Schools shouldn't ban anything unless it was purposely intended to incite violence, not just because it was expressing a personal viewpoint. And what is the legal and moral precedent for allowing the teaching and practice of Islam in some schools, yet banning the same from the Bible?[42] The only explanation for this is the lack of courage and/or conviction on the part of those who run our schools.

As I said before, one of the scariest things for a teacher is to realize that if anything bad happens in class, you will have 15-30 witnesses that could testify against you. Think about that. Your word vs. 15-30 witnesses. I remember one time when a group of kids in my class didn't like some reasonable decision that I had made and were openly conspiring to get me in trouble. To the rescue came Rhonda, an African-American girl, and one of those great character kids that would do the right thing no matter how many stood against her. I had her in my class her freshman year, and she was a junior now. She heard what the kids were saying and immediately piped up and said, "Don't worry Mr. C, I got your back." I knew that a good word from her would mean more to the school principal than a bad word from ten other kids. She had that kind of reputation. I will always remember her for that incident, because the matter did go to the principal. The potential for kids to gang up on you like this may be one of the reasons some teachers choose the path of least resistance by making friends with the students instead of challenging them. The other reason is because some teachers simply want to be liked and remembered that way. Well, I wanted to be liked too, but there are limits. I remember walking across campus one day during the first year that I taught, and running into a group of boys, none of whom I had in any of my classes. They asked me if I was Mr. C. I said yes, and asked why they wanted to know. They said they had heard about me and that everyone said that I was a real cool teacher. I felt good about this for awhile until I wondered why the kids thought I was cool. The last thing I wanted to be was one of those cool teachers whose students didn't learn anything. In that respect, the remark worried me. I had made up my mind that I would show as much love and respect to the students that I could, but I would never compromise on three things: (1) Students will get the grade they earn, (2) Instruction comes before popularity, and (3) Disrespect would not be tolerated. This automatically meant

Set Our Children Free

that I would be unpopular with some students, but I didn't care. I was there to teach, and the good kids understood that and respected me. If I was liked, OK. If I wasn't, then at least I know I did my job.

I was never afraid of any student despite my age and size. Something about old age and treachery will always defeat youth and inexperience. Besides, I was still pretty athletic, and worked out and ran on a semi-regular basis. And I had learned karate from an Okinawan master many years ago when I was stationed there with Uncle Sam. Still, some students carried a lot of anger, and when you got a group of them together, especially when they were willing to lie, it could amount to trouble. This is one reason that it was absolutely essential that a teacher carry liability insurance, unless they belonged to the union, which I wasn't about to join. I only had one parent threaten to sue me, but I've had plenty of other confrontations involving parents, student threats, and fighting that could have gotten me into legal trouble. In addition, when I first started teaching, I was told never to be alone with a student. That's a great idea on paper, but in the real world what do you do if a student comes in for help? I always kept the door to my classroom open in such instances, but what if you did all the right things and a student still lied anyway and accused you of wrongdoing? Can you prove a negative, i.e. that you were never alone with them and that your door was always open? It was easier to just spend the money on the insurance and have the peace of mind.

I recall one incident during my first year when several students had set me up to be assaulted. I had gotten into the habit of giving a 5 minute "lesson on life" to begin each class period. In one such lesson, I had made a comment about a crime I had read about in the local newspaper, and apparently one of the individuals involved in the crime was friends with several of my students. The word had gotten around that I had said something bad about him, so these students conspired to get me. As the students filed in and sat down to begin my first class after lunch, several of them told me that another kid, whose name I did not recognize, wanted to speak with me outside the classroom. Since I was getting ready to start class and I couldn't leave the students alone, I told them to tell him to come into the class and I would speak with him there. They began to insist that I go out, and I was equally insistent that he come in. When they realized that their ambush wouldn't work, although I didn't recognize it as an ambush at the time, the boy came in, got right in my face, and said something about how I had disrespected his friend. It was obvious that he was very angry and ready to fight. There was a glaze in his eyes and I believe he was probably stoned. I calmly walked over to the speaker button on the wall, asked that the school resource officer come to my room. I then told the kid that he needed to leave immediately, which he did reluctantly and angrily. I was never afraid of him or any other student during my teaching tenure, although I probably should have been. I found out later that this particular kid had a reputation as a very tough street fighter. He was expelled for threatening me, but a few years later during a similar incident with a different principal, (see the story at the beginning of this chapter) the student received only two days suspension.

There's bad behavior, and then there's mischievous behavior that just makes you laugh. Sometimes kids would break the rules but do it in such a way that you just couldn't be upset. If a kid had a sense of humor, it usually would take him a long way and get a lot of stuff overlooked. One such kid was Tanner. He and his older sister Karen were from a prominent family in the community and they both went to Hancock H.S. He had a tough act to follow, as she was all-everything – beautiful, talented (she sang like an angel), and smart (near the top of her class). Tanner was talented and smart, but in a twisted genius sort of way. I'm sure he had a very high IQ and could get acceptable grades without doing much work. He had no interest in being a student, however, even though he had a great imagination. He and a few friends showed me some of their homemade videos that they had posted on their web site. The videos were kind of a cross between Candid Camera and a B movie starring Jim Carrey at his worst. For high school students they were very creative. I couldn't get Tanner to do any work, so predictably his grades during his senior year were lower than he needed to get into the prestigious university of his choice. So his mom wrote to all of his teachers a week or two before finals asking us to make any effort we could to cut him a break, give him extra work, or anything else that could raise his grades. Of course it was too late by then. My understanding is that he got into this college anyway by enrolling during the summer session. If you do well during the summer session, you can bypass the strict entrance requirements, I'm told, and the university will automatically send you an application for the fall semester. This was vintage Tanner – finding a painless shortcut around the rules. His biggest splash however came during the homecoming celebration one year, when he disrobed down to his underwear during one of the skits and started dancing on stage. Several teachers and principals tried in vain to get Tanner down from the stage, but he just kept on dancing. When they finally managed to chase him from the stage, they couldn't catch him as he ran around the track at the football stadium in full view of the crowd. All of this took several minutes as everyone was in stitches. This was very embarrassing to the school principals since many community members and alumni were present for the event. One of these principals knew something about embarrassing homecomings, as their own daughter during a previous homecoming threw condoms from the parade floats instead of the candy everyone else was throwing. I felt sorry for Tanner's parents as they were conservative values-oriented people with good reputations. I got to meet them during the parent-teacher conference that followed Tanner's suspension. I don't know what happened to Tanner, but I'm sure that he is making lots of money somewhere and smiling.

So what's the answer to discipline problems in the school? I discuss this further in the last chapter, but the solution is clearly to have tough rules and be willing to enforce them. I also think we should bring back alternative schools for the worst offenders, and run those schools like a boot camp. They can earn their way back to a regular school with good behavior, but if they get kicked out of the alternative school for bad behavior, hand them back to their parents to home school them or put them to work.

CHAPTER 5

THE PROFESSION OF TEACHING

Teaching is an honorable profession, and that should never be minimized. My generally negative tone throughout this book about teachers, administrators, public schools, and the educational establishment does not detract from my admiration of the job teachers do on a day to day basis. My problem is with the policies that the teachers' unions and the educational establishment espouse. But when the classroom door is closed, there are flesh and blood human beings behind that door battling to make a difference in the hearts and minds of kids, just like I was, and I applaud that. Teaching is the hardest job I've ever had, and in addition to working as an engineer, I've worked in steel mills, junk yards, construction sites, and on the waterfront. I've been everything from a welder to a door-to-door salesman. But I have never had a job as challenging as teaching school. The time spent in the classroom is the most hectic and emotionally exhausting eight hours I have ever spent on a consistent basis, and I have worked on multimillion dollar projects where every minute counts. And it's not just my opinion, as surveys usually rank teaching at or near the top of the most stressful professions.[43]

To those who have never had the experience, it is really difficult to describe a day in the life of a teacher. If you are not an organized person who can multitask, you have no chance. And I don't know of any teacher that doesn't bring work home. Even on a night when you don't bring work home, the job never leaves you. Your mind is continually thinking about a lesson or incident the previous day, or what you are going to do the following day. Your students become a part of you, and you cannot push them from your mind, even when you are home. This is why most teachers prefer to be married to teachers, because sharing these problems with your mate is the kind of therapy that you can't get from a non-teacher. A teacher gets just three breaks during the school day – a 20 minute lunch, a planning period (every other day with block scheduling), and after school. If a student walks in during one of these times, for extra help or just to talk, no teacher I know will refuse them, even though it cuts into their valuable time. On rare occasions, when I could not afford to see a walk-in student, I would hide and do my planning somewhere else on campus. Although it was hard to do planning without my computer, it was easier than telling a student I didn't have time for them.

As a teacher, you are isolated to a degree that I've never experienced on any other job. All day long you are on an island with 15 to 30 other individuals, who would rate collectively on the maturity meter lower than you alone. Teachers seek each other out during the brief lunch and duty periods, since this is the only time to connect with another adult to regain some sanity and solace. The other teachers are fellow employees, but you don't work "with" them in the sense that you do in any other job I know, and they are just as isolated as you. You have a supervisor, which is the principal, but it is not close supervision since you rarely see them or talk with them. So there you are – a one person

Set Our Children Free

company . In a sense, you are your own boss, the decision-maker for every activity you are involved in, with 125 – 150 "employees" (students) to whom you have been entrusted. You will eat, sleep, and agonize over every decision, event, and lesson plan you share with these employees 24 hours a day for 10 months a year. Their successes will be your successes, and their problems will be your problems. It is a tremendous responsibility because your employees don't produce a product – they are the product. It is a job fraught by a continual undercurrent of stress, interrupted by occasional, albeit infrequent moments of intangible rewards. It is definitely not for the faint of heart.

A typical day starts with picking up the mail at your mailbox. On any given day, the mail contains countless amounts of correspondence and directives from the principal, requests for information from everyone from the school board to state agencies, notes from parents and students, surveys to fill out, and announcements about meetings and other things you are expected to know. If you don't get to school an hour before class starts, you won't have it all sorted out by the time the bell rings for the first class. In addition, you may have morning duty, which is a 20 minute period where you stand somewhere on campus hoping your presence deters students from doing something illegal, immoral, or both. Sometimes your duty period is at lunch or after school. Many teachers try to steal some extra time from these periods by planning or grading papers, etc. while standing on duty, although they can be held liable if something happens to a student while they were distracted doing work. Anyway, by the time you get the paperwork sorted out, and I never did, the bell would ring for first period and here came the students, ready or not. If for any reason you are not ready, there's no scarier feeling than standing in front of 25 high school teenagers, who expect you to entertain them, and who, despite various stages of sobriety and somnolence, are still perceptive enough to tell if you're faking it.

Then came one of the hardest parts of the day for me – taking attendance. Don't laugh. This was an extremely important activity and other teachers insisted that I was making it harder than it was, but I never did get it right 100% of the time. School funding depends upon proper counting of students. Parents are notified by auto-dial if their kid is absent. Excused vs. unexcused absences can mean the difference between passing and failing a class to a student. Yet I couldn't seem to get it right. In fact, my first year of teaching, I didn't even take roll. At that time, the attendance was supposed to be kept manually in your grade book, but if someone told me to do it, I don't remember. When I told two of my fellow teachers near the end of my first year that I hadn't taken attendance all year, they laughed until they cried. They couldn't believe that a teacher had gone an entire year without taking roll, and that I could somehow have thought that it was that unimportant. "I know who's missing," I would say, and they would laugh some more. They were right, of course. It was VERY important, but the engineer in me tended to eliminate what I considered frivolous details and boil down the job to its bare essence – instruction. That's why I ignored a lot of the other paperwork I got in the mail also. I just didn't have time for it because there were so many details to take care of, and that

Set Our Children Free

first year I was anything but organized. I figured that if some piece of paperwork was important, someone would let me know about it sooner or later, and they did, sometimes impatiently. I remember Dr. Sanders, our principal, telling me one time that I wasn't being very professional about the way I handled some of these details. She was right. Some teachers let their students take roll, and I tried it for awhile until I discovered that they didn't do it right either, and some of them actually cheated by not marking their friends absent. After my first year, they changed to a scanning system where the scan sheets had to be returned to the school office where they were processed, and an automatic dialer called the parents when the student was absent. This had to be done within a certain amount of time after the period began, but here's where it gets complicated. Some of the students didn't show up for class until after the scan sheets made it to the office. If they didn't have a late pass, they then had to be marked either tardy or absent depending upon how late they were, and the sheet had to be corrected. Then there were other students who had excused absences because of a field trip or whatever, and they could not be forgotten and had to be accounted for also. This means you had to keep track of the field trip permission slip which the student likely gave you two days ago. Good luck with that! To complicate matters further, some students would show up for class, then immediately go to the restroom, so you would miss them while taking roll. The school locks the restrooms before and between classes, so the students have to wait until they get to class to go to the restroom, and most of them can't wait until class gets started. Also, other students would come in and out of the classroom for various reasons, with late passes from another teacher, or the office, etc. And you had to do all this at the beginning of class while other students were coming up to your desk, distracting you by asking you any number of things like, "Can I go to Ms. Butler's class for a minute to get my backpack?," or "What's my grade in this class?," or "There's gum on my seat, can I change?" Combine these distractions with my antipathy towards trivial detail anyway, and taking roll was a Chinese puzzle I never seemed to figure out. Occasionally, I would interrupt myself right in the middle of class and look at a student I knew I had marked absent. Of course, they would tell me they had been there the entire time, but somehow I had missed them while taking attendance. At the end of the semester, there would always be a student or two that would insist that I had recorded too many unexcused absences for them, and it had to be corrected or they would fail. I would look on my computer, and if I had them absent on a particular day that they said they were present, I would usually give them the benefit of the doubt, especially if it was more than a week earlier. I knew my limitations.

This is why I also had very little regard for another useless exercise teachers were required to endure – lesson plans. Teachers are required to submit to the school administrators a description of each lesson they were going to teach during the following weeks. Some administrators required a great deal of detail in these lesson plans, causing you to waste precious planning time on a purely administrative requirement that added nothing of value to classroom instruction. Once a teacher submitted their lesson plans, they disappeared, never to be seen again. I seriously doubt any administrator even read them, and they were never even filed as far as I could determine. Therefore, in my mind, this was simply a bureaucratic function that existed for the sole purpose of assuring someone that teachers received adequate oversight - a myth in itself. If teachers are honest, most would admit that they have no idea what they are going to teach a week from now, as so many factors can change their schedule. Besides, I think a teacher needs to be flexible instead of feeling pressured to keep a rigid schedule that may or may not be in the best interests of the student. Having no patience for this kind of nonsense, my lesson plans rarely reflected what I actually taught on a given day. I wasted as little time as possible on them, sometimes just throwing something together that looked like a plan, and I never got challenged on them, which told me they never got read.

Following that first hectic five minutes of class, there was a great deal of flexibility allowed as far as what a teacher could do in those opening minutes of instruction time. I took full advantage of that by bringing up subjects for discussion with the students that had nothing to do with my curriculum on math or science. I did this to get the students to think about things outside the box - that went against the conventional wisdom of their friends, the news media, their favorite TV shows, or even their own school. These were cultural or values-oriented discussions, and no subject was off-limits. I was up front with my principal about the fact that I did this, and she assured me that anything that was reported in the newspaper was fair game. That was all the license I needed. To me, this was a great opportunity to engage the kids in a subject that was usually of interest to them while allowing me to pass along some values that could help them in life. But the first day of any school year was very important, and a student would usually remember everything about that day even if they remembered nothing else that school year. Knowing this, I asked myself what I would talk about if I only had one day with these students. The answer came in a small poster that the school handed out, which I attached to my wall in the front of the room. It was an American flag superimposed with this portion of the Declaration of Independence: *"We hold these truths to be self-evident, that all men are created equal, that they are endowed <u>by their Creator with certain unalienable Rights</u>, that among these are Life, Liberty and the pursuit of Happiness. — That <u>to secure these rights, Governments are instituted among Men, deriving their just powers from the consent of the governed</u>..."* Since there are no lessons or books on the first day of school, and most of the day is spent going over the class rules, syllabus, etc., I used this poster to generate a discussion about the meaning of liberty in our country. I had the three phrases underlined, just like I did above, pointing out that, 1) our rights come from our Creator, not government, 2) the only purpose for government to exist is to secure those rights given to us by our Creator, and 3) those just powers which the government does possess, come from we, the people. It was amazing to see that most of these high school students had never heard or even thought about these concepts before.

Following this discussion period, was to me the easiest and most enjoyable part of teaching in my opinion – instruction.

Block scheduling made it harder, of course, because the classes were 1 1/2 hours long instead of the normal 45-55 minutes, but the actual passing on of knowledge is the part I enjoyed the most. Now, in order to be an effective instructor, I had to deal with many obstacles, and this is the part I enjoyed the least. The obstacles were, first of all, block scheduling, which meant you had to stretch the normal student attention span of 15 seconds to more than an hour and a half. In order to accomplish this, you had to change the method of instruction at least twice during that time period, and throw in something of entertainment value if possible. If you didn't, students would throw the "B" word at you: "This is **Boring**!" It would always be said in a tone similar to what they would use if they were accusing you of rape, i.e. "How dare you violate my mind with stuff this boring!" I would usually explain that if I was an entertainer, I would be in Hollywood, not here. Or I would say that many things in life, from learning to making a living, are boring activities that we still have to find a way to do because they are necessary. Even when I was their age, I understood that. But this concept was so alien to them that I might as well have been talking in a foreign language. Another obstacle to effective instruction, and by far the biggest, was student misbehavior. I spend an entire chapter on this in another part of the book, but let me briefly say that this can be the most frustrating part of the teacher's day. If there were no behavior problems, teaching would probably be one of the most rewarding jobs in the world. If parents and taxpayers knew how much time, creative energy, and talent was wasted on dealing with student discipline, they would demand change. On many occasions, these behavior problems suck the life out of a school, and the joy out of teaching. I rate this as one of the top two problems afflicting our educational system today. Fortunately, teachers rarely have all bad classes. There are usually one or two good classes to which they can look forward. When the day is over, you either pick up the pieces and try not to make the same mistakes the next day, or build on your success if you had a good day.

If you are a teacher, you get sick time, but you hate using it even when you're in a body cast. The reason is because you have to get a substitute. There is nothing wrong with substitutes per se', as they are good, well-meaning people. It's just that nothing will get done on any day that you miss, so it is a wasted school day, which I will explain. Because you don't know ahead of time which day you will turn up sick, you have to prepare an emergency lesson plan. This emergency lesson plan must be, by design, a generic "busy work" type of lesson because you never know what time of the year you will wake up sick, and thus which part of the course for which you need to plan a lesson. So if you wake up sick on any given day, and have to stay home from school, your substitute will have to use your emergency lesson plan and the day will be wasted. Even with planned absences, when you know which day you'll be absent, the kids will never do the assigned lesson because, to them, a substitute day is a day off – a "free" day. Even if you have the substitute require them to hand in their work for a grade, half of them still won't do it. The kids feel entitled to a day off when you're not there, and many times even ask for a "free" day when you are there. This is why it is even harder to be a substitute teacher than a regular teacher. On a substitute day, there are more behavior problems, classes skipped, and cheating than a normal day. In addition, when you know that you will need a substitute, you never know who you will get. It may be anyone from an ex-military drill sergeant who makes the students march in formation and say "yes sir," to a flower child who makes the students sit in a circle, hold hands, and talk about world peace. And if that isn't enough to worry about, you have to put everything away and lock it up tight when you need a sub, because if you don't, everything in your room will either be destroyed or stolen by the time you get back. I am not exaggerating. I've gone as far as unplugging my computer, display screen, keyboard, and even the mouse, and locking it up in the back room on a day when I knew I would be gone. As a science teacher, this was particularly worrisome, because you always had all kinds of gadgets and displays out on the tables demonstrating some scientific principle. If you didn't put them away, it would be the last class you ever got to use them. Now you know why teachers would rather have a root canal than miss school.

There were few places to have lunch without being interrupted, and one of them was the infamous teacher's lounge. The first thing I want to do is dispel the notion that the teacher's lounge is a smoke-filled gossip club where liberal philosophy flourishes and where student reputations go to die. It's not true – they don't allow smoking anymore. Other than that, it's a pretty accurate description. But the price you pay escaping to the teacher's lounge, was having to listen to some of the teacher conversations. Understand that for a fiercely independent soul like me, it was torture sometimes to listen to some leftist nonsense and keep my mouth shut. When I didn't, a polite argument usually ensued, and I was usually in the minority. There were only a few teachers in the school who tended to see eye to eye with me, but they didn't always have the same lunch schedule as me. So if one of my like-minded friends didn't show up, I ended up talking with someone else, and the conversations usually ended with one of them saying, "You sound like a **Republican**!" I need to take a moment here to adequately describe the inflection with which that phrase was uttered. Picture equal amounts of both shock and disgust. Shock, because most teachers have similar views and are not used to being confronted with a point of view other than their own. And disgust because the thought of another teacher having such views and possibly passing them on to the next generation of young people was just too sickening for them to contemplate. The tone of their voice made it sound as if they had just discovered that a deadly virus had broken out on campus. My reply was always the same – "I'm not a Republican - those guys are **way** too liberal for me! "Republicans too liberal?" they would snort. The shock and disgust had now turned to panic, and by this time they were dialing the Department of Education to get my teaching certificate revoked, and if possible having me detained under the Patriot Act.

At least I wasn't a student. The teacher's lounge can be a very cruel place for students, as they are the No. 1 topic of gossip. As a teacher, you should not be biased toward any student based upon what another teacher says about them, but you're only human. And it works both ways. Students whose teachers say good things about them will generally get the benefit of the doubt from other teachers. But students who are slackers or give a few teachers a hard time will get no breaks from the other teachers. Is this fair? Of course not. A student may have a personality conflict with a teacher, and that teacher only, or they just may not like that particular subject. But they may be an angel in every other class and do their work well, and still not get the benefit of the doubt because of a thoughtless word from a frustrated teacher. Studies have been done on elementary school children, where teachers were told lies about the achievement potential of a group of students.[44] In actuality, all of the students in the study were chosen at random. The students who were identified to the teacher as high achievers, did in fact show dramatic increases in their test scores, due to the teacher's predetermined expectations of them. Does anyone think that the reverse would not be true? I tried not to let what other teachers say prejudice me, but I can't say for certain to this day that it didn't have some effect on my attitude towards a particular student. I tried to withhold judgment until that student did something that confirmed what I had been told. It certainly didn't affect their grade, because I was always very clinical about the way I graded – students simply got what they deserved whether I liked them or not. Only one time, and one time only, did I violate this principle. I had a kid named Andy in one of my science classes that was a typical lazy lower tier student. However, he was a nice kid who was always respectful to me, but didn't do much work in class. I received regular calls from his mother, who like Andy was very respectful to me, and tried to work with me on getting him an education. Finally, the mother asked if she could come and sit in my classroom to make sure Andy did his work. Most teachers would have refused this request, but I thought she not only had a right to do it, but I welcomed it. This was a working mom, and I thought that anyone who loved their son that much should be given the benefit of the doubt, even at the risk of embarrassing herself and him. She did this more than once. At the end of the year, after final exams, Andy was still nearly two points below passing, with a 58.5% average. I called him up to my desk, and showed him his average, and then had him watch as I changed it to a 60% - a passing grade, which allowed him to at least get a credit for the course. I told him that he deserved to fail, but that I was going to pass him solely because of what his mom, and not he, had done. I then told him that he needed to go home and tell his mom how much he loved her, for being the kind of mom every student should have.

Along with the low pay and the hectic day comes lack of respect. And one of the reasons for the lack of respect *is* the low pay. I taught in a state where teachers couldn't strike so the pay was even lower than many other states. However, I had little sympathy for teachers who complained about the salary when they knew what the scale was before they even began college. Still, I figured that if the teachers' unions would be open to letting the free market work, then supply and demand would take care of the problem, at least for some teachers in critical shortage areas. There has been a critical shortage of teachers in math, science, and special education in my state for a long time now, but there is an

Set Our Children Free

easy and responsive way to address it. The local school boards, during negotiations with the teachers' union, should simply ask for the power to offer better pay for those teachers who are in higher demand. They don't do this because it would ignite a holy war, since equal pay for all teachers, good or bad, is an untouchable union tenet. So it is left up to the state legislature to address the issue, and in many states there isn't the political will to do so. Our state has made it easier for people in non-educational technical fields to transition to teaching without taking that many extra college credits, but more needs to be done. It doesn't make sense for most aspiring teachers to pursue a difficult major in math or science if they want to teach, when they could go into a technical field with just a few extra credits and make a lot more money. I know many people who did just that. One of our science teachers went back to school and got a few extra credits, and now works as a pharmacist making three times her teaching salary. If school boards could offer market rate pay to attract teachers, there would be no teacher shortages in any subject area. Because they can't, they are left to fend for teachers from other school districts or other states, or try to convince people in other professions to switch careers and go into teaching. There is a problem with both of those strategies. School boards want to attract teachers from other districts and states, but they don't like to give them the same number of years of experience they had in their old job, which means the teacher would be essentially starting over towards their retirement. And as for attracting people from other professions, they need to somewhat make up the gap in pay, so that the financial penalty is not so great. I will discuss solutions in more detail in the last chapter of the book, but believe me, I know there are no simple answers to raising teacher pay in general, even though it can be done for some. My wife was on a school board for 12 years, and I know how sensitive it can be when trying to balance the needs of teachers against the interests of taxpayers.

Given the above, I don't think low pay is the main factor for lack of respect, but it is a factor. I didn't realize it until I taught for a couple of years and noticed how many times the issue was raised by both parents and students. They would question my competence and knowledge because they knew that I didn't make that much money. Sometimes the implication was subtle, but sometimes they would come right out and say the following quote verbatim: "If you know so much, what are you doing here?" Students have been programmed by our society to believe that a person is only to be valued if they make a lot of money. And it was clear to me that their parents felt the same way during any number of conversations I had with them. Because I wasn't making big bucks, I must be stupid or foolish, and therefore not worthy of respect. Students never failed to mention the fact that they drove a nicer car than me (I drove a truck or motorcycle), had a nicer cell phone (they usually did), and would be making more than me when they graduated. I got a little more respect from students when they found out that I was teaching because I wanted to, not because I had to, but then I would usually hear, "Why would you put up with all this crap if you didn't have to?" Their value system had no room in it for those who gave to a cause higher than themselves, regardless of the rewards. Such people were fools, in their opinion, because there was only value in acquiring things.

And the disrespect doesn't stop with the students. The teacher is the face of the educational system, for good or evil, so they generally receive the brunt of any dissatisfaction the public and parents feel. As teachers, we have to endure lots of abuse and insults from students, and you can write that off to immaturity. But most of the time, if you have a disrespectful student, you also have a disrespectful parent. That makes for some very interesting phone calls home, and some very eventful parent-teacher conferences. Because most kids lie to their parents about anything bad, from low grades to discipline problems, teachers have to constantly document everything to protect themselves. Even then you could get in trouble. As a teacher, you have to control your temper and bite your tongue so many times when kids and/or parents are yelling, cursing, and threatening you. That's part of being a professional, but it isn't fun and it isn't part of my nature. Parents who act this way because they are foolish enough to believe their kid's lies, and haven't a clue as to why their kid is failing, despite all your attempts to help them, are particularly disturbing. What I really want to say to them during those times is: ***"Stop wasting my valuable time with your spoiled rotten, low-life, lying, cheating, morally reprehensible kid, who would have to be stuck with a needle full of concentrated ambition just to be considered lazy! And take their illiterate, ill-begotten, inconsiderate, lard-butt out of my class, so a more deserving student can get an education, and while you're at it, get them sterilized so they can't reproduce, and that goes*** *for you too! If by any chance, you are reading this, and you're one of those parents who used to yell at me about your kid, realize that I'm not a teacher anymore, and if you EVER talk like that to me again, I'll tell you what I REALLY think of you and your kid!!"* There. Now I feel better. **LEGAL DISCLAIMER:** *Any person or persons, student or scholar, whether real or imagined, having a description exactly, or similar to, the above described "kid" and bearing any resemblance accidental or coincidental to person or persons living, dead, presumed dead, or asleep in my classroom on or about the time period when I used to teach school, during which time period said person or persons above attended the same school and classroom as the one I taught, is purely coincidental.*

Maybe I should have just taken the advice of one of our wise teachers, Joe, with whom I talked with on a regular basis. He said he would tell parents that he would make a deal with them - If they believed half of what their kid told them about him, he would believe half of what their kid told him about them. Got that? Whether the parent understood it or not, it usually shut them up.

This brings me to the main source of disrespect for teachers – they have no real authority. I won't dwell on this here, since I discuss it thoroughly in my Chapter 4, but unlike a generation ago, kids have no reason to fear teachers. If a student is disrespectful, a teacher can't paddle them, can't fail them, can't beat them to a pulp, and murdering them is definitely frowned upon.   And absent a few loose cannons like myself who banish disrespectful kids from their classrooms, there is no good alternative for dealing with disrespect. All a teacher can do is notify the offender's parents or write them up and hope their school administrators will do something to deter any such future behavior. Good luck with that! A call to a parent of the disrespectful kid will likely get a step-parent/guardian who has no control over them because the kid doesn't respect them either. A write-up will result in a slap on the wrist at best – a one-day in-school suspension where they can hang out with their peers and high-five each other over the way they cussed out a certain teacher. Kids are disrespectful to teachers because there are no meaningful consequences for doing so – it's that simple.

The vast majority of our faculty was trained from day one to be teachers, i.e. they had education degrees. I only knew of one other faculty member that had spent a significant amount of time working at a job in the private sector, and he was kind of a maverick like me. We were largely untainted by the prevailing theories about education. I had to eventually know these theories because in my state, we were allowed to teach certain physical sciences and math with a technical degree but we had to complete a half dozen college credits in "teaching" courses within a few years of the time we started. In addition, like any other teacher, we had to pass the state's subject area examinations for our area of certification (math and physics in my case), and had to pass the state teacher's certification exam. If we did all of the above, we would get a teacher's certificate, good for five years, in our subject of expertise, and it was only good for primary, middle school, or high school, whichever specific area for which we had tested. While the subject area exam is a fair objective test of knowledge that any teacher teaching that subject should know, the teacher's certification exam is loaded with subjective questions that are student oriented. You are given a series of situations involving teacher and student. Many of my colleagues didn't know how to answer the questions, but like myself, they were told to answer every question from the perspective of <u>what would be perceived as best for the student</u>. It worked, illustrating that the prevailing theories of education are defined from the student's point of view, since that was what they were obviously testing. Depending on the subject, there were additional certifications required, at your own time and expense, of course. For instance, if you taught English, our state required that you take 300 hours of ESOL (English for Speakers of Other Languages) training, due to the high Hispanic population, and to settle a federal civil rights lawsuit against our state. Even as a math and science teacher, I was required to take 60 hours of ESOL training, which I never did, partly because I was told it wouldn't be an issue unless we were audited, and partly because I thought it was another stupid federal mandate that I didn't want to spend 60 hours of my precious time and money on.

Since I did not come from an education background, I believed then, as I believe now, that it gave me an advantage. By an advantage, I mean that it gave me a unique perspective to all that I saw going on around me. I had not learned the prevailing liberal theories about how to teach and handle students. All I knew about teaching was what I had learned from my teachers during my formative years a generation earlier, as well as some of the instruction I had done in Junior Achievement and youth groups. I admit that I was "old school" in my approach to a lot of things, but again, I don't consider that a disadvantage. I remember being frustrated my first year while telling Dr. Sanders that I felt somewhat like an albatross on my campus. That didn't disqualify me in her eyes, but it would ultimately be my undoing as a teacher in the eyes of a different principal a few years later. In that respect, career wise, it was a disadvantage. I still believe that you can't change who you are in order to teach. If you're conservative and old school, it's going to come out sooner or later. If you're liberal, ditto. It may be partly responsible for why fewer and fewer men go into teaching, and more of them leave teaching once they've been there. Men are typically more conservative than their female counterparts. It's gotten so that male teachers are nearly nonexistent in primary schools, and they're becoming scarce in middle and high schools also. I feel that one reason is because they are less able to adapt to the disrespect, permissiveness, and political correctness that exist in today's schools. They typically teach with a more authoritative style and tolerate less nonsense and disrespect, yet that is exactly what they are expected to tolerate in today's school environment. I know of no male teacher that is happy with this trend, and if it continues, male teachers will be an endangered species. I know for a fact that many of the ones who remain are only doing so either because they love to coach, or because they have too many years toward retirement to quit. We need male teachers not only as role models for kids who don't have one, but to communicate the male worldview to the student population. I do not intend to demean the female worldview, but at the moment it is the prevailing philosophy not only in our schools, but in our national policymaking as well, and we could use some balance. I could write a book on this subject alone, but my purpose is simply to point out that everything in our society from government domestic policy and foreign policy, to movies and TV has been feminized. The administration of our schools is no exception.

At the school where I taught, only about half the teachers belonged to a teacher's union. The union did provide aggressive representation for teachers who were unjustly accused of something, which happens quite often. Unfortunately, they also provided aggressive representation for teachers who were justly accused of something, and should have been suspended or even dismissed. From what I saw and heard, most of the teachers who joined the union did so because they felt they needed job protection. There was no question that it was harder to get rid of a teacher if they were a union member. While the union provided protection for a teacher who didn't deserve it, it also represented many teachers who did deserve it. In many school districts, where there is strong union representation, it can cost well into six figures and many years of litigation just to fire teachers guilty of the even the most egregious acts, let alone incompetence.[45] Our district was not that bad, but there is no question that any teacher needs legal protection, and in fact, they would be crazy to go without it. I personally never joined the union but I did have legal representation, which I came close to needing several times. One teacher in our county was accused by a student of sexual misconduct and fought the allegations all the way to the school board. He was exonerated by the board and cleared of the criminal charges as well, and to my knowledge is still teaching today, albeit with the stigma of this charge over his head. Without legal insurance, he would have faced a monumental legal bill. My cousin left teaching over a similar untrue allegation. One of our principals was sued for getting too aggressive in breaking up a fight. I could give you dozens more examples of frivolous lawsuits and allegations. So why wouldn't all teachers join the union? Well, the union dues were steep in my opinion, but for some, the union's politics were a little too hard to swallow. The most prominent teachers' union supports a far left political agenda that would shock most Americans. And the school boards accommodate them by providing automatic deductions from teachers' paychecks, which significantly aids their ability to respond quickly to non-politically correct uprisings, which most of us would call common sense reforms. Those states that have blocked automatic deductions from members' paychecks have seen union membership and dues drop dramatically.[46] When our local union head approached me about joining the union, I asked him if I could deduct that portion of my dues that was earmarked for political activity. He just smiled and walked away, with an expression that seemed to say, "It's irritating having to deal with knowledgeable people." The biggest kept secret in America today is that the Supreme Court has ruled that dues paying non-union members can't be compelled to pay for political activities with which they disagree, because it is a violation of free speech.[47]

Set Our Children Free

A welcome trend to me has been the move to hold schools accountable, via student test scores, for meeting certain standards. This has been vigorously opposed by both the unions and educational elites, but it makes absolute sense. There should be no disagreement that schools need to be held accountable for teaching at least the basic principles of a core curriculum, or they have no purpose in being. And, just like a teacher has to test their students in order to assess whether they've learned anything, schools must likewise be assessed and graded to assure that kids are making adequate yearly progress. If this seems like common sense to you, then your instincts are correct. But it just goes to show how much of a disconnect exists between the goals of most parents/taxpayers and the goals of the educational establishment. Educators may rightfully debate about details such as how money is distributed under the law, but trust me, it is the accountability to produce results that really sticks in their craw. During the latest round of legislation to hold the schools accountable in my state, the TV and radio ads were rife with spots declaring the end of education as we know it. The bill also eliminated teacher tenure. One teacher was quoted in the newspaper as wondering how he could focus on his students' education when he didn't know if he would be employed from one year to the next. I would like to have told him, "Welcome to the world that the rest of us have to live in! In the private sector you have to produce something of value, or you won't be employed the following year, whether you can focus on your job or not." My personal experience is that most teachers would teach the same, with or without tenure, because most are professionals who like doing what they do. But I've often wondered why a free market system that has worked extraordinarily well in private industry, where we only produce durable goods, has never been allowed a fair chance to work in an educational system where we produce something far more valuable.

One of the nice things about being a teacher at the end of the school year was the little things you get to do to leave a legacy with a student, particularly seniors. Many would ask for a letter of recommendation to submit with a college application or student aid request. I usually took the time to write an excellent letter because of the weight it carries with college administrators. Seldom would a poor student ask for a letter of recommendation, and if they did it was usually for a job, not college. I never refused one, but I was never dishonest about what I wrote – I mean every student has **some** good things I could say about them. Another joy was signing a yearbook, because what I wrote would be there for many years. In addition to writing about my best memories of the students, I would always include some guidance, and I've always felt that the best guidance came from Scripture. Most of the time I included one of my favorites passages from the Bible, Proverbs 3:5, 6 - "Trust in the LORD with all your heart and do not lean on your own understanding; in all your ways acknowledge Him, and He will make your paths straight."[48] Some students would also write very touching words to me on their senior school photo, or in a separate letter or card. When students take the time to let you know how much you've helped them, it can erase all of the bad moments in an instant. It is these moments that we remember forever.

Set Our Children Free

Another joy is seeing former students return after they've graduated, or others you just run into later on. Some of the most memorable reunions were from students who were now in the military service. A former student named Ned walked in one day and related to me how he scored high on his Army entrance test because of the stuff I had gone over in my Principles of Technology class. I also received a wonderful letter and a subsequent visit from Amanda for the same reason, and she was now working in Army intelligence. Kerry was another former student who stopped by and was excited about her career in the Navy. Other students kept in touch by e-mail, including one of my foreign exchange students. Most of them stop communicating after awhile, and I understand that perfectly. A teacher can be an important part of their life, but a less important part with each year that goes by. Teachers can never be friends, in my opinion, but we can be mentors, confidants, and role models. Normally, the former students who seek you out are ones that have fond memories of you, and vice versa, and I never had any trouble remembering these students or their names. On the other hand, there are the ones you just happen to run into by chance. You usually have forgotten their names, particularly if they never did anything to distinguish themselves. You then have to either fake it, or risk embarrassing them by asking their name and apologizing for your brain freeze. This has happened more often than not when I encountered one of my former students, even if I recognized their face. One of the most notable examples I can think of was when I was vacationing in Daytona Beach, which was at least a 2 ½ hour drive from our school. I was walking out to the beach from the hotel after dark when I heard someone shouting my name from at least 50 yards away. The kid ran up to me and told me he was in my class a couple of years earlier for just one semester. He told me his name and remarked that I was a "helluva teacher." We talked for a few minutes and had a nice conversation before parting ways and wishing each other well. Just one thing bothered me. I had no recollection of ever seeing this kid before in my life, and his name didn't even sound familiar. And what are the odds, I thought, of running into one of my former students that far from home, and that he would recognize me from that far away in the dark? I mean I rarely run into one of my former students anyway, and I only live in the adjacent county to the one in which I taught. I've always found it strange that it only took me two or three classes into a new semester with my students to remember all their names, and yet I could forget those same names so easily once I stopped seeing them every day. I actually knew teachers that admitted that they went an entire semester and still didn't know all of the kids' names. This was not uncommon, by the way, as other kids complained to me about teachers like this. It made me wonder how they could care so little to not even memorize their student's names, and how they managed to even give those same students a grade.

Then there is the downside of losing kids who will never graduate or ever return to see you again. Tragically, we lost an average of two students per year, most of them from car accidents. It is particularly shocking because you don't expect young people to die, at least not at the rate of two per year in a student population of only 1400. When I attended a high school of similar size, it seems that there were far fewer fatalities, and in fact I can only remember one during my high school years. And yet Hancock was not unusual in that regard, as most neighboring schools had similar figures. And if such a tragedy struck one of the students in your class, you never forget it. The most painful loss I can remember was Jared, a happy-go-lucky kid on our soccer team. I had met his mother and father at open house, and could tell that they were solid citizens who provided Jared with a good values-oriented upbringing. Jared and another student, Mike, were out one afternoon enjoying themselves, when their jet skis stalled, leaving them stranded in the cold water of the Gulf (it was winter in Florida) near sunset. An all-night search ensued, but they died of exposure before they could be found, and had drifted many miles from their last known location. Jared's father gave a courageous and moving eulogy for his son at the funeral, and it was so difficult to see the pain that they were going through. I kept Jared's seat empty as a classroom memorial to him the remainder of the school year. Mike was from a good family also, and I had previously had his sister Kelly in one of my classes. There were other students who I had previously taught, who died in subsequent school years, and yet others who were the brother or sister of a student I taught or coached. Several of them were good kids from good families, and it is particularly heartbreaking when you know the family, although in a small town it seems as if everyone is related at times like these. I would tell the parents that I couldn't explain why bad things sometimes happen to good people. I could only pray that God would somehow help them cope and someday make sense of the loss of their child.

I once saw the passage below posted in a teacher workroom, and I have never found out who the author is, but I guarantee it was a teacher. So since I can't credit the tortured soul who wrote this, I will dedicate it to all of the teachers out there who are overworked and underappreciated. Only someone like me, who has walked in your shoes, can truly know what you go through each day:

"Let me see if I've got this right. You want me to go into that room with all those kids, and fill their every waking moment with a love for learning. Not only that, I am to instill a sense of pride in their ethnicity, behaviorally modify disruptive behavior, and observe them for signs of abuse, drugs, and T-shirt messages. I am to fight the war on drugs and sexually transmitted diseases, check their backpacks for guns and raise their self-esteem. I am to teach them patriotism, good citizenship, sportsmanship and fair play, how and where to register to vote, how to balance a checkbook and how to apply for a job."

"But, I am never to ask if they are in this country illegally. I am to check their heads occasionally for lice, maintain a safe environment, recognize signs of potential antisocial behavior, offer advice, write letters of recommendation for student employment and scholarships, and encourage respect for the cultural diversity of others. And, oh yes, teach, always making sure that I give the girls in my class fifty percent of my attention."

"I am required by my contract to be working, on my own time, summer and evenings and at my own expense towards additional certification, advanced certification and a master's degree. I am to sponsor the cheerleaders or the sophomore class, and after school, I am to attend committee and faculty meetings and participate in staff development training to maintain my current certification and employment status. I am to collect data and maintain all records to support and document our building's progress in the selected state mandated program to "assess and upgrade educational excellence in the public schools."

"I am to be a paragon of virtue larger than life, such that my very presence will awe my students into being obedient and respectful of authority. I am to pledge allegiance to supporting family values, a return to the basics, and my current administration. I am to incorporate technology into the learning, but monitor all web sites for appropriateness while providing a personal one-on-one relationship with each student. I am to decide who might be potentially dangerous and/or liable to commit crimes in school or who is possibly being abused, and I can be sent to jail for not mentioning these suspicions to those in authority. I am to make sure ALL students pass the state and federally mandated testing and all classes whether or not they attend school on a regular basis or complete any of the work assigned."

"I am to communicate frequently with each student's parent by letter, phone, newsletter, and grade card. I am to do all of this with just a piece of chalk, a computer, a few books, a

bulletin board, a 45 minute or less plan time, and a big smile on a starting salary that qualifies my family for food stamps in many states. Is that all?"

*"You want me to do all of this, and **YOU EXPECT ME TO DO IT WITHOUT PRAYING?"***

CHAPTER 6

PUBLIC VS. PRIVATE VS. HOME SCHOOLING

I was actually sorry to leave public school, despite the way I was treated. I had invested several years in the teaching profession and I liked it. I was also looking forward to coaching a great group of kids on the tennis team to a fourth straight district championship. I applied to a couple of other public schools but my heart wasn't really in it. I felt that the principals at my school would give me a bad reference anyway since I did not leave on good terms. It was clear to me that I was never going to be their kind of teacher. I felt they rigged the evaluation process in my case, because they knew I would leave voluntarily rather than submit to their way of doing things, and I did. In my mind, it was so hypocritical of them to give me excellent marks for instruction (they had little choice, since I was a good instructor), and bad marks for "teamwork" (Translation: we can't get you to lower your standards to our level. There is no teamwork in the teaching profession, since you never work with other teachers, and you are on an island 100% of the time.) Not being a "team player" meant that I was a loose cannon, which I admit. It was a tough job, but somebody had to do it. Strange that the previous school administrators didn't have a problem with me, and gave me good marks. Not that they agreed with me all the time either, but at least they recognized my value and acknowledged that I was much more of an asset than a liability. To be honest, I wasn't sure that any public high school principal would ever see eye to eye with me on the educational issues I considered important. That's when I thought I would give it one more try at a local private faith-based school, which I'll call Hillside Academy, in the same town.

I had a good feeling about Hillside the moment I walked on campus. You could sense a peace and spirit of cooperation about the place. Parents and kids seemed happy to be there. I had a great interview with Mr. Hannity, the Superintendent. He said all the right things about their philosophy of education and how they ran their school. We saw eye to eye on all of the important issues, or so it seemed at the moment. He said they believed in high standards, lots of homework, school discipline, and parental involvement. In fact, they had a Web-based system of communication for parents much more advanced than Hancock had. And a biggie - they had eight periods per day instead of the abominable, detestable 3-period block scheduling. Because their teacher salaries were even lower than those in the public schools, and because I was certified in the critical shortage areas of both math and science, Mr. Hannity offered me an extra $5000 supplement – something public schools were not permitted to do. I would also be utilized as a coach in one or more sports. As I walked around the campus, I felt I had found a home. Their facilities were every bit as good as the public school where I had taught, they shared my view of education, they valued their teachers as well as the students, and everyone seemed to enjoy working there and being part of the team. The only down side was that they didn't offer health insurance, so I had to purchase my own policy, which wasn't a big deal since I was in good health and had passed a physical with flying colors.

My first semester at Hillside, I thought I was in heaven. The kids were wonderful and well-behaved. The staff was courteous and helpful. The coaching went great. I had made some new friends among the faculty. The school administrators, Mr. Hannity and Jack, the principal, seemed to be great character people and good leaders. This was the way school should be, I thought. I couldn't get over the contrast between public schools, where everything was out of control and hedonistic, and this school where order and good values reigned. There were a couple of seemingly insignificant things that should have raised a red flag in my mind that first semester, but didn't because I was having such a good time. One red flag was that I had to teach Chemistry, which was the only non-math course on my schedule. Well, I didn't have to, but when Jack said it would make things easier schedule-wise if I would, I volunteered because I wanted to be a "team player." I really didn't want to teach Chemistry. It was the one science class that I never really wanted to teach, but I was stuck with it. It would eventually prove to be my downfall at Hillside. Another red flag was that I kept hearing how the students at Hillside were spoiled and had driven other teachers off because they had a reputation for wanting A's without working for them. This didn't seem to be a problem for me, since I taught as I normally did and nobody complained. I had a great relationship with most of the students. The third red flag was the two in-service teacher professional days where Mr. Hannity and Jack spoke to us about grade creep. They said

Set Our Children Free

the grades at the school on average were way too high, and that we needed to change the culture and challenge the kids. They repeatedly emphasized that they would back us if the kids and parents complained and put the heat on. Again, this didn't bother me because I continued doing what I always did, and no one had complained. The combination of these three red flags should have alerted me. Little did I know that I had one semester left in my teaching career.

Christmas vacation was wonderful that year because I felt very much accepted as part of a family of faculty and staff at the school. One of my student's parents threw a great Christmas party at their very expansive ranch outside town. We also had everyone over to my place for a Super Bowl party where my hometown Steelers won the game. There was one other pleasant surprise that Christmas season when bonus checks were handed out. I wasn't expecting anything because it wasn't announced, but Mr. Hannity was a very effective fund raiser for the school. He told the faculty and staff that he didn't think the bonuses would be as much as last year until an anonymous donor came through. I heard the staff gasp when he announced that the bonuses would be the same as last year. Since I didn't know what they were the year before, I thought the check he handed me would be a token $100 or so. I mean, how much could a small school like Hillside afford to spend on Christmas bonuses? When I opened the envelope I understood the gasping. It was a check for $2000! Counting the number of staff at the school and doing some quick math, I figured that someone must have given the school a gift of about $60,000. This was a welcome gift, because my finances definitely suffered during the years I taught school. It was a sacrifice I was willing to make, but it was sure nice to be working at a school where teachers were truly appreciated.

There wasn't a single negative thing I can remember about that first semester at Hillside. I enjoyed teaching the classes, and the team I was coaching was doing well. There was a discipline problem now and then, but they were mild compared to public school. The staff and administration were very supportive also. In public school most of the peer pressure kids felt was to <u>misbehave,</u> be it sex, drugs, fighting, or disrespect. At Hillside, most of the peer pressure was to <u>behave</u>. I got to know a lot of the kids there, and trust me, they weren't all angels. In fact, I estimate that about 50% or more of them were there because their parents made them go, not because they bought into the particular religious teachings of the school. They were normal in every way, very similar to the kids in public school. Then why was there little if any misbehavior? The reasons are twofold, I believe. First, as I said earlier, there is tremendous pressure to conform to the rules, unlike public schools where anything goes. If you took some of the Hillside kids and put them in public schools, I'm certain that they would react just like their public school peers. The second reason is equally simple – bad behavior was simply not tolerated. Students knew that disrespect towards a teacher and/or repeated violations of the rules would result in expulsion not just a slap on the wrist. Public school students have no such fear because the worst that could happen to them would be a ten day out-of-school suspension, and only for major crimes such as drug or weapon violations. One of the fundamental psychological tenets is that all behavior is a result of its consequences. Kids knew that few, if any consequences followed misbehavior at the public school, but quite the opposite at the private school. But, you may argue, we can't kick kids out of public schools. Their parents' taxes paid for the schools, and everyone has a right to a public education. Actually they don't. Hillside's parents paid school taxes too, but they had to pay an additional tuition, which in most cases far exceeded their school taxes. And last time I looked, public education wasn't one of the rights enumerated in our Constitution. And even if you did make the case that it is a right, then I could make an equal argument that keeping some of the bad kids that we tolerate in public schools, robs other good kids of their right to an education. My opinion is that there should be no compulsory education past the eighth grade or age 14. If a kid has no interest in school, or is not succeeding, or is a troublemaker, there is no point in going further. It shouldn't be up to the school system to babysit a teenager. They are old enough to get a job, or complete their schooling online at home. If we force them to attend through high school, they will be a drain on the time and resources of teachers and kids who really deserve an education. Remember that compulsory school attendance laws are a relatively recent phenomenon in this country, and were not at all the rule in many states until the 20th century. Even now, only about half the states require attendance until age 16.[49] Both of my immigrant parents had to drop out of school by age 16 and work to support their families, but they were far better educated than the average 18 year old high school graduate today.

Set Our Children Free

The kids at Hillside were good enough that near the end of the year I let my guard down and failed to store my final exam in a locked file cabinet as I would have done at Hancock. The door to my room was locked so I figured that would be enough. When I graded the final exams I was surprised to see Zeke and one other student, neither of whom was known for their intellectual prowess or study habits, score in the 90's. Zeke was a graduating senior, athlete, and a student at the school every year since kindergarten. He got the highest grade in the class on the final exam and he attributed it to studying hard. I was suspicious, but I couldn't prove anything. Zeke had looked me right in the eye and swore to me that he didn't cheat. One thing that's always true, however, is that kids can't keep a secret, especially when they think they got away with something. Word soon got around to several of my top students that Zeke had broken into my room through the window and made off with a copy of the final exam. They told on Zeke because they were upset that he had cheated his way to a better grade than them. Unfortunately, Zeke had graduated early and been given his diploma already. Mr. Hannity said something about trying to change his transcripts but it never got done. I was told not long ago that Zeke had changed and is now a church pastor, just a few years out of high school.

One of the few discipline problems I had was a kid named Wally. I had been warned that he had a very protective mother, and how she could make a teacher's life miserable. Wally wasn't really a bad kid in the sense that public school kids were bad. He was just immature, talked a lot in class, and did some of the little things that tweak a teacher into insanity. I was no exception. Wally and his mom did to me what they had done to others, although it was nothing malicious. Apparently when his mom heard I was leaving at the end of the year, he told me she was going to make up a T-shirt which had a picture of Mario (my nickname) on the front which said, "I survived Mr. Caruso" along with the current school year. He thought I would be mad at the idea, but instead I suggested that she have a bunch of them made up, bring them to school, and sell them as a fundraiser. He did and they sold well. I still have mine. Mario is the name students at Hancock gave me and it stuck even at Hillside. Apparently they figured because I was Italian and had a mustache, that I looked like the cartoon character on the Nintendo Super Mario Brothers games.

Just as at Hancock, there were students at Hillside that seemed to gravitate towards me, and they would come in and feel the need to talk on a regular basis. Two that come to mind were Adrian and Katrina. Adrian was a former runaway who was now living with her older sister. She had a love for Beatles music and the 60's classics, a taste which surprisingly many of the kids of the current era share. She was a pretty blonde girl who always seemed to be caged in by the environment there, and out of place at Hillside – a free spirit who stayed within the rules only because she had to. She occasionally made several remarks of a sexual nature to me, which surprised me, given where we were. She was smart enough to get average grades without working, and so she did little work. But she always had a smile on her face, and an attitude that led you to believe she had her sights on something bigger and better beyond the school years. Katrina was the face of the school, the school's poster girl if you will. She was athletic and gifted musically as well as academically. She was the perfect teenager, or so it seemed. Katrina was involved in everything at the school and was especially busy because it was her senior year. She took to me immediately and asked if she could come in and talk to me after the very first class. We soon found that we had a lot in common, surprisingly. She came several times a week after that and we often had lunch together in my room. After a few talks, it became apparent that what looked like the perfect life on the outside, was a very troubled girl on the inside. She felt her parents were unduly strict on her and she resented them for it. It wasn't just the fact that she couldn't date or have most of the freedoms other kids her age took for granted. It was the way her parents treated her, in her eyes, that deeply discouraged her. According to her, they never gave her credit for anything, and never appreciated her enough for her accomplishments as a talented musician, athlete, and Christian. All she ever got was negative talk and criticism, and they felt she was slipping spiritually. Whenever she strained at her restrictions, she was accused of going down the wrong path, and wanting the wrong things in life. I didn't know her parents, but whether her stories were true or not, I felt very sorry for her. I had heard these kinds of stories before from some of the public school kids I had in my classes. I would observe them and think that they were the kinds of kids any parent would be proud to have, and then later find out that their parents never seemed to be happy with them. What a shame, I thought, that these parents were such perfectionists that they couldn't appreciate what they had. By the end of the school year, Katrina was calling me a good friend. She had gotten a job, an inexpensive car, and a boyfriend, whom she had to sneak around to see. We communicated once or twice after she left school and started college at a school a few hours away. I remember her telling me she used a bike for transportation, had gotten another job, and a tattoo – something to which I'm sure her parents would have strenuously objected. I never heard from her after that, but I still pray for her today as I do for several other students with whom I had close relationships. In her case, I was very afraid that she would go astray from her core beliefs when she left home because of the rebellion that had built up inside her. I hope that my fears were unfounded.

Sometime into the second semester Mr. Hannity submitted his resignation, effective at the end of that school year, because he said he felt called to be a church pastor. Not long after that Jack also decided that he had some other ministry in mind and was going to leave as soon as he found out what and where it was. I was sorry these gentlemen were leaving because I felt, at that time, they were good leaders and the school would miss them. I would soon wish they had never worked there in the first place. The warnings and the red flags that I had ignored about some of the students had started to materialize that second semester. I noted earlier that kids never work as hard the second semester, and that they really slack off near the end of the school year. Hillside kids were no exception. Add to that the well-deserved reputation they had for wanting something for nothing, and you had the recipe for an uprising. It started, as it usually does, with a small group of spoiled girls, who decided that the Chemistry class that I didn't want to teach in the first place was way too hard. Forget the fact that they had gotten good grades the first semester. They stopped studying altogether, and as their grades dropped they convinced several other students to sympathize with them. They soon had significant minority of students organized against me. They focused their criticism on one particular test that most of the class failed. I wasn't too concerned about that test because it comprised a very small percentage of their total grade in the class. The top students were still getting A's, but some of the complainers were in danger of getting C's, and a few even (horrors) D's for their quarterly grade. For the top students, this test was pretty much their only blemish for the year, but for the newly-formed "posse" it was now a convenient excuse for them to do no further studying, and thus they failed several more tests after that. By now the parents were panicking and calling the school and demanding answers. They were also demanding that the test be thrown out, or the kids be allowed to re-take it. When Jack came to see me, his biggest fear seemed to be that some of the kids would flunk, not for the year mind you, but for the third quarter. I assured him that if he waited just a week until the third quarter grades came in, that I didn't see any of the kids flunking. He replied, shouting, "We haven't got a week!" I thought, wow, the pressure must really be getting to him. He didn't force me to change any grades, although it was clear that he desperately wanted me to, so after thinking about it, I decided to throw out the test in question. I knew what would happen though and it turned out exactly as I thought it would. The group of kids whose grades were in danger now saw this as a win, and an excuse to make no further effort the remainder of the year. I gave them a re-test and they failed that one also just as I thought they would. Their parents had the principal and superintendent on the run and they knew it.

The criticism spilled over into another class, as Jack told me he was now hearing from Neil's mom, someone he "never hears from." The implication was that since Neil's mom had never complained about any other teacher, then it must be my fault. Parents love to use this reasoning against teachers. "My precious Johnny never had a problem with any other teacher but you." When I got this from parents, I was always tempted to say, "Yeah, but I don't have any problem with any of the other kids but yours." Of course, when you ask the other teachers about their precious Johnny, you get a completely different story than the parents told. Most of the time teachers just take the path of least resistance and just give in. Neil was one of the laziest students I ever met. When I assigned classwork, he never made any effort to do it, even when I offered to help him with it. He had no interest in learning. He simply expected to be passed. For this, I had the pleasure of getting chewed out by his mom on the phone and berated unmercifully. She simply would not accept that Neil had any responsibility for his work whatsoever. And of course I had to be very nice and acquiescent while she was yelling at me, like all good teachers are supposed to do. It comes with the low pay.

The third quarter grades averaged out just like I thought they would - very few C's and maybe one or two D's, and no failures. In fact, for the year, two thirds of all of the grades I gave were either A's or B's. That means only one third of all my students received a C or below, and I believe I may have only failed one student – Neil, but then again, he richly deserved it. Those grades are well above the typical Bell curve average, and certainly no evidence of a teacher who grades too hard. One of my fellow teachers said that he heard one of the top students from my Chemistry class complain that this was the first time since he had been at the school that he actually had to work for a grade. "I think that's great," the teacher told me. He knew how spoiled the kids had become, and how many teachers this same senior class had driven off because of similar behavior.

The die was cast and the damage was done however. Remember how I said that Mr. Hannity and Jack held two meetings with the teachers earlier in the year, complete with charts and graphs, asking us to change the culture and do something about grade creep? And remember how they said they would back us up when the parents put the heat on? Turns out when the heat was on, they folded like a cheap suit. Well it so happens that the end of the third quarter is the time every year that they renew the teachers' contracts for those who are coming back the following year. They would put a contract in each teacher's mailbox and ask them to sign it if they intended to return. I didn't get one, and I wasn't concerned, but I did wonder why. A couple of days later Mr. Hannity walked into my room while I was having lunch and said he wanted to explain why I didn't get a new contract. "We decided to wait until the end of the year before we make a decision on you," he said. Given all that had happened the previous three weeks, I wasn't surprised, and yet something about what he said or didn't say bothered me. First of all, he didn't explain <u>why</u> they didn't give me a new contract; he just said they wanted to wait until the end of the year. I tried to draw it out of him, but he clammed up and wouldn't talk. Finally I told him that I was under the impression, based on what he told us earlier in the year, that we were trying to change the culture at the school. I think this caught him by surprise, the fact that he was experiencing a hypocritical moment, so he hesitated before giving me his one-word answer - "Eventually." I wanted to say, "Funny, you didn't say 'eventually' earlier in the year when you declared the War on Grade Creep. The message we got was to change the culture now." He said nothing but those two sentences before he left, and this was strange behavior from a normally very talkative man. One of the other school administrators told me later that they never gave him a reason why they weren't renewing my contract either. In fact, he told me he thought I was the best math teacher they ever had at that school.

I rarely act in haste, but the more I sat there and thought about it, the more steamed I became. I loved that school and I had worked hard all year to make it a better place. I was certified in both math and science, had real-world experience in the applications of those subjects, was a darn good coach and role model, and was financially in a position to work long hours for low pay. Not to boast, but who else with my qualifications would be willing to do that? I frankly had too much pride to wait until the end of the year for them to make up their mind. It was important to me to leave on my own terms. The way I felt was that the school's management had the better part of a year to evaluate whether I was an asset or not, and if they hadn't seen enough reasons to keep me by now, then I was wasting my time. I had no more to give. By the end of my next class I knew what I had to do, and there was no point sleeping on it. In my mind, I had no alternative. I typed out a quick, but respectful resignation, effective at the end of the school year, and left a copy on both Jack's and Mr. Hannity's desk, since they weren't in. Although there was one complete quarter left in the school year, neither man ever discussed my resignation or even acknowledged that they got my letter, which told me it was the right move on my part. I let them off the hook. They never had to muster up the courage to come and explain to me why they weren't retaining me. If they were really in doubt about whether they wanted me to stay, they would have come to me after they read my resignation letter and tried to persuade me to give it some thought. And even if they wanted me to go, I thought it strange that they never talked with me about it again, or even offered thanks at the end of the year for some of the good things I had done there. I thought it showed a genuine lack of class on the part of both men. To not even say goodbye or thanks was bush league. If I was that bad of a teacher, then why even let me finish out the year? They could have fired me at the end of the third quarter when everyone was complaining. At least I had the character to finish the year instead of quitting like I could have. Mr. Hannity never did land that pastorate, and since he had already submitted his resignation, he had to take another job as a principal at a different school instead. I'm told that a couple of years later when a Hillside team visited his new school to play an away game, Mr. Hannity was very cold towards the Hillside contingent. In my mind, that confirmed the point I made earlier.

Some of my fellow teachers were not happy that I was leaving and voiced their displeasure. They had seen this happen before and knew what the school was losing. I was sorry also, since I loved the kids, and had made some good friends among the faculty. One of the hardest things I had to do with my students was to gently tip toe around the subject of why I wasn't coming back the following year. I couldn't just come right out and say, "Our school administration didn't live up to their word, so they made me a sacrificial lamb to a few indulgent parents." But I couldn't blame myself either because I hadn't done anything wrong. And if I just said I wasn't coming back without giving a reason, the kids would assume I didn't like teaching there or teaching them. As you can imagine, this put me in a difficult position. I couldn't tell them the entire story until the end of the year, which really hurt. Surprisingly, when I did tell them, they understood. They had seen it before. I had a good rapport with 90% of the sophomore and junior students, and even more than half of that infamous senior class. One by one they came to say goodbye at the end of the year, many of them tearfully. They showed a lot more class than the school administration.

I went back to Hillside once the following year for a basketball game to see some of the kids I had coached. I also ran into many of my other students and they seemed genuinely happy to see me, and the feeling was mutual. I can truly say that I miss most of them and think about them to this day. There are many things not to like about teaching. It is hard, self-sacrificing work, done many times for students, parents, and school administrators who don't appreciate it. But it all seems worthwhile for the memories of those special relationships that you forged, and the hope that what you gave to that relationship made them a better person. This is why, despite all the reasons to hate it, we still love teaching.

So that was the end of my teaching career. Oh I was offered a job unsolicited from a principal at another private school who said he agreed with my philosophy of education, but I knew the moment I submitted my resignation at Hillside that I would never go back. Most people would look at the financial sacrifice I made while teaching and would see my return to engineering as a no-brainer, given that my income tax bite alone was more than my gross annual teachers' salary. However, some things cannot be measured in monetary terms. I was never particularly motivated by money to begin with, as my parents always stressed that there are more important things in life. I am happy doing what I am doing now, but call it a fault; I would still go back to teaching today under the right circumstances. Those circumstances would be that I could do it my way – the right way. Since that's not going to happen any time soon, I am left to educate others about how our schools need to change.

So what are my conclusions about private vs. public education? As I said earlier, the kids are the same, but the rules at private schools both constrain and liberate students to a much greater degree than they do at public schools. This gives private schools a huge advantage, as they do not have the burden of government regulation suffocating them. They have much greater flexibility in every aspect of education from curriculum to discipline. Therefore the kids learn more, are better behaved, and get the benefit of religious/moral instruction in an atmosphere conducive to learning. For this reason alone, private schools are a 100% improvement over public schools. The disadvantages? Cost, of course, is usually beyond the means of the average parent, especially since they are already paying more than enough taxes to send their kids to public schools. Give them back their tax money and allow them to spend it on whichever school they choose, and public schools would shrink by at least 50%. But cost is not the only disadvantage. When a parent spends a lot of money to send their child to a private school, their expectations sometimes become unrealistic. They are not willing to settle for underachievement either in or out of the classroom. They expect high grades because of smaller class sizes, more individual attention, and the emphasis some schools place on college preparation. And in many cases, they expect a change in their child's behavior, because of the more disciplined environment and/or spiritual atmosphere. This can lead to the kids themselves believing that they are privileged, and entitled to receive good grades whether they make the effort or not. On balance, however, I think that any parent who has a choice between sending their kids to either public or private school would be foolish not to choose the latter. There is that much difference, and the advantages far outweigh the disadvantages.

Set Our Children Free

There is another alternative that has surfaced in recent years, and that is charter schools. Charter schools are actually public schools that are chartered by a state or local school board, depending upon the law of that particular state. Every state's charter school law is different. They are chartered for a certain number of years by contract, and there is usually a cap on the number of schools. They receive public funding, although typically not as much as regular public schools, but are free from many of the rules and regulations that apply to public schools. Students attend by choice, and there is usually a waiting list, as these schools have become more and more popular, particularly in the inner cities. Charter schools also can offer specialized education in particular curriculum areas, or make any other innovations that generally aren't available to regular public schools. They can even cater to certain types of students. In exchange for this freedom, charter schools are held accountable to the same state academic standards, or their charter could be revoked even before the end of the contract period for which the school is chartered. They can be managed by government agencies, private companies, non-profit foundations, or other entities, and can receive private funding also. As you can imagine, teachers' unions are not particularly supportive of these schools, since they don't have the same restrictions on hiring, firing, pay, etc., and are free to hold teachers accountable. Predictably, union and government studies tend to show that charter school students underperform regular public school students, while private studies tend to show the opposite. The largest charter school experiment in the nation, however, is in New Orleans, which since Hurricane Katrina in 2005, has a majority of their student population in charter schools. From 2005 to 2009, the percentage of failing schools there has dropped precipitously, while virtually all of their top performing schools were charter schools.[50] Surprised? I'm not, and the evidence there is overwhelming and irrefutable.

Charter schools are a welcome trend, but it should only be a start. States should go much further in several areas. They should do away with all rules and regulations except the requirement that their schools meet the same academic performance requirements as other public schools. Unfortunately, states and school boards currently hold charter schools to even higher standard, as they will close down charter schools for performance that they routinely tolerate in other public schools. One of the reasons many charter schools fail is because the funding is so meager that the schools have a hard time furnishing facilities, transportation, and supplies. Charter schools usually have to rent buildings and provide their own transportation, even though they get less money from the state. Lawmakers should assure that per-student funding is equal to the regular public schools. Also, the time in which teachers spend teaching in charter schools in most cases doesn't count towards their retirement in the state retirement system, even though they are public schools. This discourages many teachers from taking jobs there, particularly if the pay is lower because of less funding. If these are truly public schools, why don't they get the same per-pupil funding that other public schools get? And why can't the teachers' time there count towards their state retirement? And why do the regular public schools get all of their facilities and transportation free, while charter schools have to scramble around for the same? The laws as they stand are simply half-baked measures borne of political compromise. Go all the way and free our schools from the monopoly of government regulation and the teachers' unions. Let every school be a charter school, and give the best facilities and campuses to the school which signs up the most students each year. Some communities have even gone so far as to make their local public school a charter school. This is a welcome trend, far too long in the making, and I believe one of our best hopes for reforming the nation's schools. Put a little competition into the equation, and the best schools will attract the most students. I believe that implementing the above, along with giving us back our tax money and letting us send our kids to the school of our choice, is the most promising avenue for reform. A recent movie, "Waiting for Superman," tells this truth on the big screen through the lives of parents and kids who have lived the hard reality of failure of our public schools, while striving for a chance at a better education at one of the top performing charter schools.[51] Students in these charter schools in disadvantaged inner city areas, far outperform their public school counterparts because they are not constrained by strangling government regulation and teachers' union contracts.[52] Teachers' unions are understandably upset about this movie, but before you dismiss this film as a right wing plot, realize that the director, Davis Guggenheim, was the same director for *An Inconvenient Truth*. Even in California, of all places, a sweeping law now in effect allows parents to fire their school administrators and teachers, and move their kids to the school of their choice.[53] Although this law is not perfect, realize that if it can happen there, it can happen anywhere, and there is hope. School choice also works well in many other countries, many of whose students score better than ours on standardized tests.[54]

Even more promising is a recent Supreme Court decision that allows states to give tax credits to private schools. The importance of the decision, **Arizona Christian School Tuition Organization v. Winn,** cannot be underestimated. The Court said that taxpayers do not have standing to sue when the state gives tax credits, (as opposed to actual taxpayer money), to private school tuition organizations. This is the first time in a generation where the Supreme Court has allowed such an exemption, and the newly formed conservative majorities in the 2011 state legislatures should not let this opportunity go to waste. It is now legal for states to give tax credits which benefit those parents who choose to send their kids to private schools, even if they are religion-based.[55]

There is only one superior choice, in my opinion, to even private schools – home schooling. Even when I taught in public schools, I would urge troubled, exasperated parents to home school their kids if at all possible. I spoke from experience. I had home schooled my son for a year and a half when I left my job in semi-retirement, and saw his performance improve by leaps and bounds. There was far less frustration and fighting with school officials, and we got to spend quality time together. He thanked me almost on a daily basis. Besides my own experience, I have since met and talked with many parents and kids involved in home schooling. The poster child for home schooling, of course, would be Tim Tebow, who excelled athletically and academically, in addition to being an exemplary human being. He went on to do the same at the University of Florida, and is now playing professional football. His experience, however, is not uncommon. At Hancock H.S., the best athlete on our varsity basketball team was a home schooled boy. The best player on our baseball team at Hillside was a home schooler also. Two of the best players on my granddaughter's softball travel team were home schooled. A friend of ours had seven children, all of whom she home schooled. I've met many other home schoolers, and they all have a quiet maturity and confidence borne from spending more time with adults than teenagers, who generally have their minds on much more trivial pursuits. In fact, I have yet to meet a home schooled kid who wasn't like that. As a whole, they score much higher than their peers on standardized tests, and adjust to college easily, a fact not lost on college recruiters.[56] And this doesn't even count the added benefits of parents spending quality time with them, nurturing the kind of relationship most parents and kids want. And since there are as many as two million kids being home schooled in our country today – a number estimated to be growing at 7-15% per year[57] - there are active home school associations in most communities where kids can connect with each other. But what about the parent who can't afford to quit their job? If you have the option of having them stay with grandparents, for instance, during the day while they do their assignments, you can complete the instruction in the evening when you get home. This is not the ideal situation, but you have to ask yourself if it is

Set Our Children Free

preferable to send them to public schools if they are not thriving or surviving there. My experience, and those of others who've home schooled, is that your child will learn twice as much in half the time. This is why home schoolers are two to three grades ahead of public school students on the average. The personal one-on-one attention, plus the lack of distractions means that you can teach them more in two hours than they will learn in public schools all day. Besides, a lot of what they do learn in public schools is not what you want them to learn anyway. Don't worry about whether you're qualified. You don't have to be a teacher or have a teaching degree in most states, and the teaching materials are readily available, of very high quality, and absent much of the political correctness that has infected public school books. Trust me, you can do this. You can get all the help you'll need from the home school associations. And now there is the additional option of having your child earn real credits from online schooling. In Florida, it's called virtual school, where a student can earn all the credits necessary for graduating without ever leaving home. I have taught all three – public, private, and home school. I can't give you any greater testimony than this: If I had to do it all over again, none of my children would have ever darkened the door of a public school for even one day. And if I have anything to say about it, my grandchildren won't either.

Chapter 7

THE TEACHER AS A COACH

They say you learn more by losing than winning, but I've never believed that. I believe that the greatest lessons in sports, as well as in life, are learned when you work hard to overcome obstacles and achieve your goals. Such was the case one night when my team played what I consider the most exciting basketball game I've ever seen at any level - high school, college, or professional. If this same game had been played in the NCAA tournament or in the NBA playoffs, it would have gone down as the greatest game in basketball history. Our opponent that night was a school from an adjacent county that had a well-deserved reputation for producing good teams, and they were intimidating just walking into the gym. They were about 20 players strong, with the normal number being closer to 11. They looked bigger, stronger, and faster than us. Even their coach was intimidating, as he was about 6'10" and still looked fit enough to be playing himself. One rule I have always followed as a coach though, is to never let my team go into a game thinking they don't have a chance to win. It was my job to not only convince them that they could win, but to devise a game plan that gave them the best chance to succeed.

From the opening tip, it was clear that this was going to be a war. This particular game had so many swings of emotion that I was convinced we had lost it many times. Our

Set Our Children Free

opponent was as good as advertised, but every time they surged ahead, it seemed to elevate our play as our kids met the challenge. What our team lacked in talent, they made up in grit and guts. Not only did our kids play their best game of the season, I was making all of the right coaching moves too (a rarity). It was punch-counterpunch the entire game. With only seconds left in the game, we were behind by three points. We got the ball to our best player Carl and he hit a 3-point shot at the buzzer to send the game into overtime. I almost expected Carl's shot to go in, because that was the level of play I had seen from both teams the entire game. Carl then promptly fouled out, leaving us with our three best starting players on the bench for what would ultimately be three overtimes. Already short on talent, it seemed hopeless without my three best players. But I could not have been prouder of our guys who were on the court at that time. I could see a change in attitude in them that was so inspiring, as they dug even deeper. They grew in confidence with each passing minute that they could really pull off this win despite the setbacks. They hustled, they dove for loose balls, they defended like their life depended on it, and they knocked down free throws like they had ice water in their veins. They functioned so well as a team that I played the same five guys for all three overtimes without making a substitution, even though they must have been near exhaustion. I was afraid that if I substituted, I would mess up the tremendous chemistry and resolve I saw from them. The other team was trash talking to our guys, which I'm sure served to motivate them even more. At the end of the second overtime, I finally began to believe we had the game won. One of my players had just sunk a free throw giving us a three point lead with

1.8 seconds to play. Our opponent would have to go the length of the court and sink a three pointer in less than two seconds – pretty much a lock, right? I ordered all of the players back to the other end of the court with instructions not to commit a foul. The other team inbounded the ball, and threw it to half-court where one of their players launched a prayer of a shot, a real Hail Mary. I cannot describe the feeling I felt as the shot hit nothing but net. Their team was going crazy celebrating, but I had to re-focus our team immediately. I mean what do you say to a team that had played their hearts out, but had just observed what seemed like Divine intervention against them? Our hometown crowd was stunned, as the moaning and unbelief I heard was almost as loud as the earlier cheers. Inside I was feeling angry that our team wasn't going to be rewarded for all of their hard work, but I didn't share that with the players. I put on my best face and told them it was OK, and that we were just going to have to play one more overtime to get the "W," even though I had serious doubts we had anything left. We did play one more, and did get the win, and even though the third overtime didn't come down to a dramatic last shot, there were some very tension-filled shots and free throws along the way. Everyone who saw or participated in that game agrees that it was one of the greatest they had ever seen. When I see one of the players from that team, we still talk about that game. In the life of a teacher, it doesn't get any better than leading a group of kids to perform beyond what even they thought possible. I just wish I could have seen these same kinds of miracles more often in the classroom.

Set Our Children Free

I have to admit that one of the reasons that I decided to teach, and one of the joys of teaching for me was the opportunity to coach. Coaching afforded me the opportunity, as a playground athlete and gym rat, to still be a kid myself. The only sport I ever played in high school was baseball, and in college it was wrestling and weightlifting, where I won my school's championship. But I had played a lot of playground basketball, and a lot of tennis and golf, and had coached softball and other sports even before I taught school. I was still in good physical condition in my 50's and my passion for sports had not waned. In fact, coaching was one of the few arenas where I felt very comfortable being a leader. The other was teaching. So when I had time, I would hang around the gym after school and get into some pick-up games or some one-on-one competition. It was always good natured and fun, and most of the time I earned a little respect in the process. One day some friendly trash talking got me into a wrestling match, literally. The wrestling team trained in the gym also, and I got into a testosterone-laden conversation with one of the wrestlers, Al, who I had in my class the year before. He was a speedster tailback on the football team and broken off a number of long touchdown runs that year. After getting permission from Coach Samuels, we went at it for a few minutes. I was leading 1-0, when Coach stopped us and barked over to Walter for tag-team reinforcement. Walter was one of his top wrestlers, and had finished 6th in the state the year before. I don't know whether Coach Samuels was embarrassed that I was beating one of his wrestlers and felt he had to teach me a lesson, or if he was impressed with my wrestling and wanted his best guy to get some work in. Regardless, Walter, like Al, was in my weight class, and we wrestled for a few more minutes with neither of us drawing blood or points. He did get a take-down right at the end when I stood up, thinking we had stopped, and I lost my concentration. Understand that I hadn't wrestled since I was in college 30 years earlier, and two minutes of wrestling is probably the most exhausting two minutes in sports. Nevertheless I was amazed that I wasn't that tired, and I felt that I was stronger than either kid. What I had lacked in technique and quickness after all those years, I made up in strength, and I think they were surprised (as was I) that they couldn't muscle the old guy around. However, I did decline to wrestle one of the other state finalists on Coach Samuels' team, who was several weight classes heavier than me, and was able to bench press about 400 pounds. This wasn't WWF and I didn't relish getting my neck broken.

I can recall my very first practice with my basketball team. I did not know any of the players well since I had none of them in any of my classes. More importantly, they didn't know me, and I could sense an undercurrent of doubt as I ran them through their paces. Maybe it was just me, but I felt as if they were looking at me wondering what this little old guy could possibly know about basketball. Leadership has to be earned at some point, even if the leader was appointed. So when practice was over, most of the guys were still hanging around, I think just to get to know me better, since we would be spending every day with each other for the next three months. I then purposely remarked that since they still seemed pretty energetic, that I might have to work them out a little more one-on-one. I really didn't think anyone would take me up on it. This was only my second year of teaching, and at that time I hadn't played any basketball in over a year, but I just couldn't resist a little dig at the guys because I knew I needed to earn a little respect. Carl, my best player, and the one I talked about earlier in the chapter, immediately took the challenge, as he was the unspoken leader on the team and would go on to average 20 points a game that season. He was full of confidence and ready to go, but since I wasn't even warmed up, I had to fake it for a little while. I told him we would start with a game of horse. The game took maybe three minutes because I don't think I missed a shot. The other guys on the team were watching this and laughing at him, and telling him I was making him look silly. This may have psyched him out a little because he should have beaten me easily one-on-one, as he was a foot taller. Because of this, I strategized that I could not let him score from short range, and that I would give him anything beyond the 3 point line. I soon found out that he could score just as easily from out there. Like I said, I didn't know these guys or their abilities at that time, or I would have guarded him differently. Despite this, and the fact that I hadn't played in so long, the game was tied at the end, and could have gone either way, as both of us had several chances to sink the winning shot. As I recall, he drained one from about 25 feet for the winning basket when I refused to believe he needed to be guarded way out there. Regardless, the point had been made, and the respect had been earned. I had no problems the rest of the season getting the players to do exactly what I asked them to do, and we had a successful year.

Set Our Children Free

As a teacher now as well as a coach, I thought it was interesting seeing the parallel between the students' athletic and academic performance. As a rule, if they were disciplined in the classroom, they were disciplined on the field or court also. If a student was a behavior problem in class, it generally carried over to the field or court. There were exceptions of course. I coached an honors student who dogged it every day in practice even though he was a very good athlete and player. He never cracked the starting lineup all season, much to the consternation of his dad, who was always on my case. A couple of years later, this same student excelled in track, which is ironic given his lack of hustle on my team. Conversely, I've seen one or two athletes who worked hard and hustled in practice, but never seemed to cut it in the classroom. Again, these were the exceptions, not the rule. I've coached at least two valedictorians and a couple of salutatorians, as well as a number of top students, and nearly all of them worked hard at being an athlete also. I remember one honors kid named Adam who was a good basketball player and excellent athlete. At barely six feet tall, he could dunk the ball, play excellent defense, run the court, and play a mean inside game. He probably had more steals than anyone on the team. Unfortunately he couldn't shoot the ball to save his life, but he would stay after practice for as long as I was willing, to work on his shot. Adam was typical. So was Carl, another honors student, who was the best player on the same team as Adam, and went on to have a stellar high school career. There was never a time when I had to question whether Carl was giving his best effort. Also typical was Randy, a three-sport athlete who didn't seem to think practice was important until I suspended him for missing. He wasn't much of a student, but he got the message and secured his place as a starter on the team. He ended up playing football for a major Division I college, but only after spending a few years at a junior college working on his grades.

My observation is that character and work ethic are generally hard-wired into a kid, whether in or out of the classroom, and they go hand in hand – if they have one, they usually have the other. An athlete can get by on pure natural ability for only so long without putting in some work, and if they don't have the character, they usually won't put in the work. We had a star basketball player named Joe one year who was ranked #5 in our state as a potential college recruit. Knowing this player's potential for self-destruction, however, our coach followed him around like a mother hen, protecting him from trouble and cutting him more than enough slack in every way. Those on campus used to refer to Joe as the coach's "son." This kid had talent coming out of his ears, but no self-discipline due to, in my opinion, no stable family life. He caught me in the gym one day after school, where I usually got involved in a pickup game, but on this day, I was just shooting by myself. He challenged me to a game of horse, and I won without him even getting past "H." Embarrassed, he immediately wanted to play another, and by now the other basketball coaches, both of whom played college ball, were observing our game. They jokingly asked if this meant I should be ranked No. 4, ahead of him. He went on to have one stellar season, but when school ended, so did his career, as he got arrested on a serious charge and I never heard from him again. I couldn't help but think of the contrast between him and another star player from our arch rival who he happened to be guarding one night. His job was to go one-on-one with this all-county player, who was known as a good student from a good family and had excellent work habits. This other player "schooled" Joe so bad that night, that it was obvious to everyone. This particular player went on to earn a full scholarship to a Division I school and starred for four years. Joe went to jail.

Another example that comes to mind is Cal, a kid on my team who could shoot the lights out from the three point line. On top of that, during the first few games, he showed a knack for being able to shoot under pressure, hitting some key free throws. He was also a very good athlete, who played wide receiver on the football team. One problem –his grades. His teachers informed me that he was failing several subjects, and not because he was stupid. He just didn't take care of business in the classroom. I took the trouble of going around to get his overdue assignments and telling him what he needed to do to remain eligible after the Christmas break. I told him I needed him on the team and made him promise me several times that he would get the assignments in. He knew without question that if he didn't, that he would be ineligible to play in any games after Christmas. He and his dad both assured me that he had a passion for the game and that he practiced his shooting every night. Despite all this, Cal never played basketball or any other sport for the school again. He didn't get even one assignment in over the Christmas holiday. I have no explanation for this other than to refer to what I said earlier in this chapter. I had him as a student when he was a junior and once again tried to talk him into playing ball, but he just didn't want to do what was required in the classroom. What a waste of good talent, as he could have been a star player. Whenever I think of untapped potential, Cal comes to mind, although I have seen this same story relived many other times with other students and athletes. Girls were not exempt from the self-destruction category either, and in some ways seemed even more prone to it if there was a guy in the picture. I can recall at least four star athletes who could have gone to college on a full-ride athletic scholarship but chose to shack up and make babies instead. Two of them earned all-state honors in their sport, and one of them may have been the best female athlete I have ever seen at any level.

There was one time when I thought I was going to need my health insurance. One of the girls in my class named Jackie was a royal pain in the butt. She had been on one of my athletic teams a couple of years earlier, and had gotten me into trouble with my team because she had a habit of missing practice. I had a near mutiny on my hands because I let her play anyway, since she had a written excuse. In addition to that, she had a bad attitude every time I took her out of the game or played her in a position she didn't want to play. In other words, I tolerated a lot from her that I shouldn't have. She was daddy's spoiled little girl, it was a simple as that. Fast forward two years, and she was now in my science class. Two years did not mellow her, and I found myself in a parent-teacher conference listening to what a bad guy I was in front of her parents. I tolerated it for awhile because I wanted her out of my class and this conference was the first step towards doing that. Finally I could only take so much without defending myself a little bit. That's what her dad was waiting for, and he didn't even try to argue with me. He was only waiting for me to open my mouth so he could close it forcefully. He was a huge guy, who was obviously into pumping iron, and would have looked right at home in a WWF or UFC advertisement. He put his hands on the table, got up slowly, and said "We can take this outside if you want!" This kind of took me by surprise since the discussion had been civil up to that point. I don't know why I didn't just stay seated and laugh at a guy trying to turn a parent-teacher conference into a combat zone. After all, I thought I had been very kind towards his daughter during the meeting. Something in me, however, triggered a righteous indignation at such behavior, similar to the way I felt when a student was being disrespectful. I raised my scrawny frame to my feet and told him that I wasn't about to be intimidated, even though I was. He replied that it wasn't intimidation, meaning he had every intention of rearranging parts of my face permanently. I was so focused on him, that somewhere in the background I didn't hear the yelling being directed at me at first. After a few seconds, I realized that my assistant principal, Mr. Wagner, was screaming at me to leave the room, ostensibly to save my life. After I left, I filled out a police report hoping to have this guy charged with assault, since as a matter of principle, I didn't think teachers should be threatened this way. The police didn't charge him but they did tell me that this man had a series of such incidents of threatening behavior all over town. My personal opinion is that guys like this spend their entire lives still trying to prove they're a man. After thinking about it for awhile, I think that he may have been holding a grudge against me from a couple of years back, when his daughter was on my team. In any event, I made the police promise me that they would warn him not to start a fight with me if he ever saw me on the street, because whatever ensued would then be his fault. Not that I was about to fight him. It's just that I would hate to have to shoot him in self-defense. The rules are different off campus.

One year I had a girl named Abby on my softball team who was very talented but was missing a lot of practices. My rule was that when you missed practice you had to have a written excuse. I previously had a bad experience (see above) where I simply accepted a player's excuses for missing practices, only to have the rest of the team resent that same player starting the game. So I kept telling Abby she needed to be at practice because I needed her on the team. By the time we had our first game she had missed about half the practices. There was no way I could start her. This prompted a call from mom, who informed me what a great player her daughter was. I agreed, but when I told her about the missed practices, she said that Abby told her she had been at the practices, and after all, where else would she be? Several games passed and the situation did not improve, which prompted another call from mom. I told her that Abby was still missing practices, and this had to stop or I could not play her. This time the tone was much worse, with a lot of yelling and screaming on her part. She also complained to the principal, Dr. Sanders, who told me that a one game suspension was punishment enough, but then Abby should play. The problem was that Abby continued to miss about one practice every week. By then, another girl had established herself as an outstanding player at one of the only positions that suited Abby's abilities, although I did occasionally play her elsewhere. She never did crack the starting lineup that year, and I got a lot of grief for it. One time in the middle of a game, a man walked over to me on the sidelines where I was coaching and asked why she wasn't playing. Not knowing who the man was, since he didn't bother to introduce himself, I couldn't divulge that information to him, but told him that I had already discussed it with the girl's mother. That didn't satisfy him, whereupon he asked again, and I gave him the same answer. He finally walked away after muttering "you're lucky, son" which I took as a veiled threat, since I knew he wasn't talking about my gambling habits. I assume he was related to Abby, but he never said.

Another season, I had a girl named Angie, who always looked all-world in practice, and was very talented. The only problem was that once the game started she looked more like all-thumbs. On top of that she played shortstop, where we couldn't tolerate a lot of errors. The situation did not improve with more practice, because after all, she looked great in practice. After a half-dozen games, my assistant coach agreed with me that we needed to make a change. We moved her to another spot in the infield, but the errors continued. Finally, we moved her to an outfield position just to keep her in the lineup. With each change of position, her attitude deteriorated right along with her mother's attitude. When the switch to the outfield was made, it was akin to a declaration of war. Her mom not only continued to give me grief, but showed up at the games with reinforcements, who would stand on the side of the stadium near our team bench and say some ugly things directed at me. It was so bad one night that even the other team's coach noticed, and found it appalling that parents would stoop that low.

Fortunately, I've had way more highs than lows with the teams I've coached over the years, and some of my fondest memories as a teacher are the victories my athletes achieved on the field. I can honestly say that it was all about them. My personal satisfaction in winning was secondary to the happiness I felt for the kids I coached when we won, and the heartache I felt for them when we lost. There were great moments like when we beat our arch rival by 30 points when they were undefeated. There was another game when we had four starters out, yet we beat another one of our rivals by 30 points – a game in which we were losing at halftime. At the half I had to re-do both our offensive and defensive strategy, and we won going away. There was the tennis team on which I was an assistant coach, which won three straight district championships. The third championship was the most satisfying because we were not expected to compete given that we had a very young team, and several of the players credited me for helping them win their final match because of the coaching I gave them between games. Then there was winning the county weightlifting championship and having the kids decorate my truck with shaving cream.

There were great disappointments too, like the one huge upset that we didn't pull off in another thrilling overtime game. We had no business even being on the court with our opponent, as they were a private academy that recruited all over the world and sent a number of players to the professional ranks. Every coach in our state knew of their well-deserved reputation. Once again, I told our kids that they should fear no one and that they could win if they did what I said. Deep down, I didn't even believe what I was saying myself. We had played a terrible away game the night before, where we had lost to an inferior opponent because we couldn't seem to throw the ball into the ocean, let alone the basket. I figured we would be tired mentally and physically, but I was not about to let the kids think that way. Another strategy I always followed as a coach was to take away the other team's best player, and force them to beat us with their lesser players. My theory was that if you forced the other team to do something they didn't like to do, that they would have to change their game plan, forcing their secondary players into roles where they weren't comfortable. I observed during warm-ups that this particular team had an outstanding big man, whose post moves were practiced and polished. I gathered my team around me and told them to bring a hard double team on this player every time the ball went into the low post. They executed it perfectly. Our double team was so good that he couldn't even pass out of the post to any cutters in the lane, shutting off the possibility of any easy layups. The only thing left was to pass the ball back outside and try to beat us from the perimeter. The strategy worked, at least for most of the game. Their big man did not score a single point all night. I heard their coach getting more and more frustrated during timeouts. Their guards kept missing three-point shots, and

Set Our Children Free

he kept telling them to keep shooting. Conversely, our team, which seemed so out of synch offensively the night before was playing their best game of the season. We were clicking on all cylinders even though we were no match talent-wise. We got a big lead which we held into halftime, and I was beginning to believe. We were fired up. At the end of the third quarter, the game was close and we were still matching them basket for basket. Finally, in the fourth quarter, their guards started to hit a few key three pointers and the game went into overtime. With only seconds left in overtime, they were ahead by one or two points, but we had the ball with one last possession. I called time-out to set up a play, but here is where their superior athletic ability paid off as they smothered us and kept us from even getting a shot off. I will always blame myself for not adjusting properly to what they were doing in the last minute or so of that game. I told the players in the locker room after the game that it was my fault we lost and that they had played valiantly. I told them that they had played like losers the night before, but that tonight they were all winners, even though the score indicated otherwise. They had done their best and had no reason to hang their heads. I had a hard time fighting back tears as I talked with them. I was so proud of them, and wanted this win for them so bad.

Another big disappointment was the final year I coached softball. It was the most talented team I've ever had, and ironically it was the only year, after many years of coaching at all levels, that one of my teams ended the season with a losing record. Sometimes too much talent is a curse because players think they're better than they are, and won't listen to you. That's what happened this particular year. For the first, and thankfully only time in all of the years I coached, I had players actually ignoring my signs. When I asked them about it later, they were honest enough to say that they didn't want to do what I was asking them to do, and then they got disrespectful about it. At least they didn't lie and say they missed the sign. I mean what did I know? I was only the coach. And then for girls who thought they knew the game better than me, they aggravated me further by continuously violating my number one maxim: Don't get called out on strikes – particularly with runners on base. With two strikes, it is the players' responsibility to either put the ball in play or at least go down swinging. Even a bad pitch out of the strike zone could be fouled off, but there was never any excuse to take close pitches when they had two strikes. I made this clear to them from the very first practice, and pretty much every practice and game after that. Still, it did not change their statue-like pose at the plate as they watched the umpire call strike three with runners on second and third at a critical point in the game. They would insist that the pitch on which they got called out, wasn't a strike, and I was equally insistent that if it was close enough for the umpire to call it a strike, it was close enough for them to swing. When nothing else worked, I finally told them that any player who violated the rule in the future would immediately go to the bench. They continued to do it, and good to my word, I continued to send them to the bench. Instead of changing their behavior, they got more rebellious and disrespectful. These girls were all moving up to the varsity team the following year, where I was slated to be the coach. I told the retiring varsity coach that I wasn't interested in the varsity coaching position, because I would have to coach these same girls another year. I never coached softball again, making my last season my worst, and those girls went on to have a predictably bad season the following year on varsity.

You would think that kids try out for sports in high school because they love the game, but I've found that it just isn't so in many cases. To some kids, it's just an athletic social club. To others, it's a stepping stone to higher status on campus. Most girls have different reasons than guys for playing sports. I've heard girls pretty much admit that they play because boys find athletic girls attractive. In fact, one of the reasons some of the girls came out for the weightlifting team was because they thought the exercises would make them look better in a bathing suit. The difference is, when game-time comes, most boys will be found putting on their game face. Girls will be found putting on their make-up. Based on my experience, I think that athletes who play high school sports purely for the love of the game are in the minority. You can tell who they are because they stand out, and usually go on to play in college or even further. Our school, like many other schools in Florida, has several alumni who played at the professional level. The kids know this, and are proud of it, and many think they will be the next pro. They enjoy the perks that go with being athletes – attention from the opposite sex, wearing their jerseys on game day, attention from the opposite sex, favors from the teachers and administration, attention from the opposite sex, getting their name in the newspaper, and attention from the opposite sex (You did read Chapter 3 where I discuss that sex is every student's major, right?). And don't discount the favoritism from teachers and administrators for athletes. I have seen it over and over again. I knew one athlete that had 42 absences, and another girl who had almost that many, and both were allowed to "work" them off by sweeping hallways, and other such work in the school the last few weeks. It was disgraceful for a student to earn a credit this way, but it was done with full approval of the school. I also knew other students who had just one absence over the limit, and yet they were prohibited from doing anything to make up the credit. Star athletes get used to this kind of treatment, which is why some of them never seem to grow up. That's why it is rare to see a mature athlete on TV, who is humble and grateful and has a sense of perspective about all that is going on around him. These kinds of athletes almost always have benefitted from the guidance of a strong family consisting of both a mother and father. In fact, over the course of many years of coaching and observing athletes, the presence of a father is the single most consistent factor I've observed in athletes with a strong character – the so-called "solid citizen" guys. And I'm convinced that the only reason many other fatherless athletes succeeded was because, at some point in their careers, they had a great coach who became a substitute father figure to them.

CHAPTER 8

POLITICAL CORRECTNESS

Let me give you my definition of political correctness (PC), as it applies to today's schools. It is a strong pressure to conform to the prevailing philosophies of federal, state, and local education officials. These philosophies encompass a very humanistic point of view where government is the center of the universe, and the nurturing and indoctrination of its children its most important function. The only morality comes not from God, but from the sacred collectivist view of conformity to group values. Therefore, the overriding goal of the system is to foster trust in this statist dogma, which includes evolution, globalism, social engineering, stewardship of the earth, a commitment to equality of results for all, and a submission of personal goals to those of the state. No education official will openly admit that these are the goals of the educational system, but that is exactly what it has become in practice. In fact, I'm sure that most school administrators and teachers are not consciously aware of the mindset that has gradually and insidiously taken over the curriculum and administration of our schools during the past couple of generations. Many of them have never even been challenged in their core beliefs, and blindly assume that everyone else feels the same way they do. This became very clear to me during my teacher's lounge conversations and professional development sessions, described in the previous chapters. Despite this, those who speak out against the established orthodoxy will do so at their own peril, since education policymakers as well as teachers' unions are adept at silencing all dissent.

There is a long line of educational "gurus" dating back nearly 100 years that have helped shape America's education policy, most notably John Dewey, a self-described socialist and father of "progressive" education. It was Dewey, in fact, who was the chief proponent of mandatory public education as a means of undermining traditional values and using the schools as an agent of social change.[58] Since other authors have documented this history thoroughly, I will not dwell on it here, as that is not the purpose of this book. What is important, however, is that much of the philosophy espoused by these educators has its roots in Marxism.[59] Lest we forget, compulsory public education is one of the ten planks of the Communist Manifesto.[60] If you read these ten planks, you would be hard pressed to find the differences between the policies of our own government, and one purposely built from the ground up as a communist government. Purposeful or not, our schools have become brainwashing centers for the political left, at every level. Concepts like liberty, self-government, personal responsibility, and free enterprise are missing in action in today's curriculum, even though they are the principles that are supposed to form the foundation of our society. An entire generation has been educated to be comfortable with egalitarianism, globalism, feminism, and environmentalism. The struggle for power between this generation of public school students, and other freedom loving individuals in our country will be the defining cultural war of our time. The

Set Our Children Free

outcome will determine the future of our country and our world. If you think I'm exaggerating, read on.

When my wife was on the school board, a number of parents began to object to some of the programs being implemented in the primary and secondary schools. These programs were ostensibly intended to combat low self-esteem, drug abuse, and other social problems, and used meditation, guided imagery, and other questionable techniques for the presumed purpose of improved academic performance and behavior modification. As a whole, they were known by various names such as Outcome Based Education (OBE) or affective education, and focused on changing behavior via values clarification, cooperative learning (group-think), and moral relativism.[61] They were basically practicing psychology on the children without their parents' permission. There were the usual assurances from the schools that they were doing nothing wrong, and that parents could opt their kids out of the program if they wished (only after they found out about it). Of course none of this was discussed with the parents before the schools decided to play doctor with the kids' minds. Their excuse was that it was *so hard* to get all of the signed permission slips from the parents. Forget the fact that they already had to get permission for far less important one-day activities, like field trips, let alone a semester- long program of psychoanalysis. Presumably the kids needed these programs because more enlightened educators determined that they knew more than the parents about child rearing, and they had to undo all of the harm that parents had done. But as upset as the parents were, and they had every right to be, it was nothing compared to the anger and vitriol that surfaced from the teachers and school administration when we tried to do away with these programs. We had no idea that this was such a sacred cow. You would have thought that we had proposed an end to reading, writing, and arithmetic – the so-called "cognitive" programs we should have been teaching instead. This became a front page topic in the local newspaper for weeks and months. People from outside the county picked up the gauntlet from both sides. I have never seen anything like the ugliness and name-calling we experienced from so-called professional "educators." They considered these programs sacred turf, not to be messed with by mere taxpaying parents. I'll never forget them saying that we were interfering with their "academic freedom" (Translation: stay out of our business, pay your taxes, and let us do whatever we want). Every evil motive you can imagine was assigned to ordinary parents who simply objected in a civil manner. My wife was vilified to a degree that is hard to describe, for daring to openly question why we were teaching such non-academic nonsense. The newspaper editorials and letters were scurrilous. They branded all those who opposed these programs as right wing religious zealots who wanted to take over the schools. It was clear that this fight was important enough to the education establishment that they were willing to pursue a scorched earth strategy to defeat the parents who merely wanted to protect their kids. Although there were plenty of parents who supported my wife's position, I don't recall a single teacher or school administrator who felt the same, including some who were our friends. It seemed that anyone who had any stake whatsoever in the schools felt threatened by the parental interference. That ought to tell you everything you need to know about public schools. The attitude seemed to be, "How dare those ignorant parents think they can tell us trained professionals what their children need!" Who do the schools and the kids belong to, anyway? Last time I looked, taxpaying parents had ownership of both, at least on paper. If you think this overreaction by educators was the exception, you are wrong. It was happening all over the country at that time.

The proposal to eliminate these programs was defeated at a contentious board meeting, 3 votes to 2. They had won the battle but they were to ultimately lose the war, at least temporarily. Because shortly after that, Governor Jeb Bush (who most teachers hated by the way) pushed through his A+ education plan in our state, requiring schools to set achievement goals and be held accountable for the results.[62] There was going to be much less time for the psychoanalysis nonsense. Teachers had to actually get back to teaching something meaningful and useful – like the stuff that kids were going to see on the state-mandated tests. But don't think that these affective education programs have gone away. I don't doubt that some of this mind-mush is still being taught behind closed doors, but teachers know that sooner or later the children will have to produce or there will be consequences for them and the school. By the way, one of the founders of these "value-free education" programs, Dr. W. R. Coulson, now admits that the programs don't work, and in fact do more harm than good, and that he owes the parents of this nation an apology.[63]

So why was this fight so important, that grown men and women would stoop to such levels? After years of thinking about it, and having been a teacher myself since then, I now understand. The reason the education establishment felt so threatened was because they really believe that they know better than you what is best for your children. They've been indoctrinated by the leftist philosophies they learned in college and in their professional development training, and of course, you haven't had the benefit of this enlightenment. They have to undo your mistakes. They don't think like you do. They don't appreciate the individual wonder and uniqueness of your child.

And then there is the subtle, but ever-present socialist indoctrination that is gradually turning our kids into passive, submissive, obedient government serfs. From the time a child enters primary school they are programmed to group-think. Individualism is discouraged, along with awarding grades based on individual achievement. Group work is celebrated and encouraged. In group work, the smartest student usually does the work while the rest of the group gets the same grade as he or she does. If there's a better way of teaching socialist philosophy, I am not aware of it. As for grades, we've pretty much done away with them in primary school. Educators feel that bad grades will make a child feel bad. Those who achieve a good grade distinguish themselves as being better than the group, and that is discouraged. Better to give everyone a "satisfactory" grade so that no child is rewarded greater than the others. Everyone gets a smiley-face sticker on their assignment or class work paper. Everyone gets a bumper sticker that says,

Set Our Children Free

"My child is a Terrific Kid at _____ Elementary." We are so concerned about nurturing them, that we forget about educating them. Funny, nobody seemed to worry about any of this when I was in primary school. We got letter and numerical grades and there was no concern that a student who got a B or a C would feel less worthy than those who got A's.

And schools are so afraid of competition now. Many high schools have even done away with awarding a valedictorian for the same reason.[64] I remember one of my primary school teachers drilling us on our math facts. She would bring two of us at a time up to the front of the room to compete side by side. She had flash cards that had addition, subtraction, multiplication, and division problems on one side and the answers on the other. She would show us the problem side of the card, and the first student that answered correctly would get the card. When she was done, the one with the most cards won. She would continue the competition until there was a final winner, who would be considered the best math student in the class. I don't remember where I finished, but it wasn't first, and I don't remember being scarred for life over it. And even after all these years, I remember the student who beat me – my neighbor Sarah, who was super smart, and whose father, an architect, designed the high school we would ultimately attend. Despite this, I turned out to be a pretty good math student, which definitely helped in engineering school. I liked the competition, and it made me want to get better so I could win the next time. Such a competition would definitely be considered a no-no in today's primary schools. I'm just getting started, so if you're not convinced, read on.

Even on the playground, games like dodge ball are discouraged because it eliminates one kid at a time until there is only one winner left. Besides, dodge ball might teach kids aggressive behavior because you have to hit someone with a ball, and that teaches, ..gasp.. *violence!* The thinking is that if we don't program the kids to be passive, they will shoot each other when they get older. This is the same philosophy that says that all fighting is bad, even when it is in self-defense. Kids are taught if we are only nice to others, they will be nice to us. This springs from the fallacious thinking that we are born pure and innocent and have to be taught evil, instead of recognizing that we, as humans, are inherently selfish and evil, and require Divine help and self-discipline to act contrary to our own nature. We can readily see the results of this naïve thinking in our own country's foreign policy. We are apologizing to other countries for America's past behavior and telling them over and over that we mean them no harm.[65] Yet for some reason they are still trying to kill us. Imagine that! To illustrate this mindset on a personal level, do you realize that every single school fight my own kids were involved in was their fault? I know because that's what I was told by the school. If my kids started it, it was their fault for picking on someone else. If they didn't start it, it was their fault because they must have said something to get the other kid mad, or they should have been passive and walked away when they were being picked on. Either way, both kids got punished because all fighting is bad, no matter who started it. Forget the fact that my kid was acting in self-defense, fighting isn't good for group control. Your individual rights don't matter. Many of you can identify with this because you have been told this same nonsense by a teacher, counselor, or principal.

Need more examples? What about how kids are constantly told they must share. I don't have a problem with sharing per se', but it is considered wrong if one kid in the class has more of something than someone else. Every child must be equal in every way. The title of the "No Child Left Behind" law says it all. Why didn't they name the law "Give Every Child Equal Opportunity to Succeed," which is what schools should be trying to do? "No Child Left Behind" implies that it's someone else's responsibility, other than the child's own efforts, to get them to succeed. "No Child Left Behind" also implies that it is unacceptable for any child to fail, and if they do, it's someone else's fault, usually the teacher. They must all pass together or we've all failed – the ultimate in group-think.

And we haven't even discussed the curriculum which is saturated with leftist propaganda. The "save the planet" type thinking is typical, because, remember, we're all in this world together. Mother Earth is the ultimate deity, and it is up to us to save her, since she is incapable of saving herself. I remember my kids coming home from primary school many years ago and "educating" me, telling me how to conserve water and how to limit greenhouse gases. That was a generation ago, and I dismissed it as harmless stuff without realizing that it was the beginning of an indoctrination crescendo. Now children are propagandized with emotional appeals about everything from saving the polar bears to saving the Obama presidency. If you think that the few videos that recently surfaced showing school children singing the praises of Obama were anomalies, think again.[66] In my little community, a Christmas program a month and a half after the 2008 election featured kids singing, "Praise be to Obama, leader of the free world." And that didn't even make the news – the local news, let alone national news. Another recent news story featured a 13 year old who was reprimanded by an art teacher for drawing an American flag with the words "God Bless America" underneath. As witnessed by other kids in the same classroom, the teacher told the 13 year old that the drawing was offensive, and then praised another student for drawing a picture of Obama. As of this writing, the teacher has not apologized, and has not been disciplined.[67] A few days earlier, four students were suspended for wearing American flags on their T-shirts on Cinco de Mayo.[68] Remember, we are a world community, and nationalistic pride has no place in it. Multiculturalism is PC ideology. Just because we only occasionally hear of these kinds of incidents, does not mean that they are the exceptions. We just don't hear of the thousands of other times each day that they happen across America.

And as for economics, you are naïve if you think students will be taught anything at all about free market principles. The school system itself is a closed door monopoly with a socialistic system of hiring and paying teachers. I can clearly remember Michele Obama telling a group of college students not to go into corporate America when they graduate.[69] We already have a significant portion of the public who have no idea where, how, and by whom our nation's standard of living was created (Answer: private industry innovation). Many who've attended public schools think all private wealth and corporate profits are evil, and that our nation's wealth came from a government printing press. Government produces nothing, and the money it does have has been confiscated from those who do produce the goods and services it represents. Business and economics classes would be the ideal place to teach students how free markets work in the United States, or rather should work. But since 95% of all teachers have never held a job in the private sector or owned a business, how will they teach something they know nothing about? To them, supply and demand means making a trip to the *supply* closet when the students *demand* more paper and pencils. What happens when public schools program an entire generation with the socialist principles we have been discussing? That generation will think, vote, and act out what they've learned, and we will no longer be the land of the free and home of the brave. Nothing less than our liberty is at stake! There has already been one big city school district caught transporting busloads of students to the polls during school hours, where they were given Democrat-only sample ballots, assisted by Democrat campaign workers once there, and then rewarded with ice cream.[70] Does anyone doubt how a large group of uninformed 18 year olds (hopefully), mentored by an overwhelmingly Democrat-registered group of teachers will vote? And that's not even taking into account the morality of using government employees to engage in partisan politics during school hours by rounding up a captive and gullible group of students who can be bribed with a half day off school and some ice cream. I might add that it has been routine for many years for teachers to openly campaign in favor of school referendums in front of their students. The American coup begins in the hearts and minds of its youth. Abraham Lincoln was prophetic in saying, "The philosophy of the classroom today, will be the philosophy of government tomorrow."[71]

The re-writing of history is another part of that coup. We can't even say "Founding Fathers" anymore because it isn't politically correct – it might offend women, so we have to say "Founders." Remember the uproar a few years back when the U.S. Department of Education funded the development of History Standards to be used in the schools?[72] These standards were so PC that most of us would barely recognize the America they portrayed. The largely leftist view of our country with all of its warts was given preeminence instead. In fact, they were so bad, that even the U.S. Senate passed a nonbinding resolution rejecting them by a vote of 99 to 1 for what they said was a failure to respect the contributions of Western civilization.[73] Well those standards haven't gone away, and neither has the desire of the leftists to rewrite history. And while the standards haven't been adopted wholesale into many of the textbooks and curriculum, they've managed to leak their way in a little at a time. However, a new assault is well underway in the form of national education standards which will require states to adopt the radical socialist and globalist view of our country.[74] This should alarm every freedom loving American, because if we fail to pass down our nation's principles and values from one generation to the next, we will soon wake up in an

Set Our Children Free

America that looks more like Cuba. In teaching our nation's history, a great deal of time should be spent on the founding documents, with the factual data coming from primary sources. Without a thorough understanding of the Declaration of Independence, the Federalist Papers, and the Constitution, how will the current crop of students know what made our country exceptional. And if they don't know that our nation is exceptional and why, they will be the first generation to voluntarily succumb to enslavement and tyranny. They need to understand the responsibilities, limits, and costs of liberty. The cost, as Thomas Jefferson so eloquently said, is "eternal vigilance."[75]

My father was an Italian immigrant, but I consider myself an American first, and an Italian second. Many of today's immigrants and those in certain political factions find their loyalty in their group identity first, with no loyalty whatsoever to their country. This is why so many immigrants as well as our own citizens, who have received a privileged education in our country, and have reaped the many benefits that our free society bestows, now hate America. I work with a Russian-born engineer who escaped from the old Soviet Union and eventually made her way to America. She cannot understand why so many here embrace socialism when she has seen first-hand how that system decays the soul of its people. She is the most patriotic American I know. I know another Turkish-born engineer who immigrated to this country alone at age 19, and worked her way through college, earning a degree in electrical engineering. She loves America and the opportunities that few, if any other countries could have afforded her. Many other American entrepreneurs who never even went to college have successfully developed products and services that have given us the highest standard of living in the world, generating millions of jobs for others in the process. And this slow, insidious, dilution of our way of life over the course of a generation has had disastrous consequences. James Madison had it right when he said, "I believe there are more instances of the abridgement of freedom of the people by gradual and silent encroachments by those in power than by violent and sudden usurpations."[76] And he further warned, "It is proper to take alarm at the first experiment on our liberties."[77] We are way past that point now. We have not been vigilant about policing the pabulum that is being served up to our kids each and every day. Whenever someone, be it a school board member or legislator, proposes the kinds of changes I'm advocating, the PC crowd, aided and abetted by the news media, paints them as right wing fanatics. It is intended to intimidate us into silence, and it has largely worked. Nobody said that freedom isn't expensive, but it's better to endure insult and indignity now, than insurrection later.

Equally upsetting is how the left has hijacked what we used to call science. It used to be that scientific principles were established by factual data and experimentation, i.e. the scientific method. Unfortunately, much of what is taught as science these days is political philosophy, not science. This "junk" science as it is appropriately labeled, has as its roots a very humanistic view of mankind and the universe. The stubborn refusal to consider any belief in a higher Being leaves scientists with few alternatives to consider. Many scientists believe that the supernatural is not rational and therefore not scientific, because the supernatural can never be seen, felt, or proven. What they've done instead is replace a seemingly unverifiable God, with an even more unverifiable theory – Darwin's Theory of Evolution - the original junk science. I disagree, however, that an intelligent Creator is unverifiable. Using the scientific method and simple statistical analysis applied to microbiology, it can be proven beyond all doubt that life could not possibly have begun by chance, as we are expected to believe with the theory of evolution. And if life didn't begin by chance, then it had to have begun by purposeful will, intent, and design. There. I've just used more of the scientific method to prove Intelligent Design that any evolutionist ever has. I gathered data, performed the mathematical experimentation, and drew logical conclusions from it. And given the extraordinarily complex nature of even a single celled life-form, and mankind's inability to produce even a simplified version of any kind of life despite advanced technology, it is perfectly logical to conclude that a Supreme Being, far more intelligent than mankind, created all life as we know it. And yet despite being force-fed this pseudo-science for an entire generation in our schools, 92% of American adults still express a belief in God,[78] including 76% of doctors,[79] contrary to the media portrayal of believers as uneducated rednecks. Seriously, you've got to believe there's a political agenda involved, when schools have taught something for an entire generation, that 92% of adults intuitively know isn't true. I am by nature a very logical human being. I don't take things by faith, including my belief in God. I have to be convinced with my brain before my heart kicks in. I have done the homework, and there is no possibility in my mind, and therefore my heart, that God doesn't exist. Zero. Nada. End of story. All the evidence is there if anyone wants to open their eyes. I don't want to belabor the point here, because there are many good scientifically sound books such as **Tornado in a Junkyard**[80] or **The Case for a Creator**,[81] that more than adequately refute evolutionary theory. And if that isn't enough for any doubters, I'm sure that any of the accomplished scientists who've done work for the Institute for Creation Research[82] could give them the honest facts. My point is that tyrannical educators in concert with the courts have silenced all discussion on this issue to the point where no alternatives, such as Intelligent Design, can even be considered. What happened to academic freedom, which was supposed to allow a free exchange of ideas and debate? The evolutionists have painted themselves into a corner where they have no choice but to establish their theory as the state sponsored religion, all the while pretending it is science. I've made this challenge to many high school and college educators as well as some media pundits, and I've never had any takers, but I will throw down the gauntlet anyway, and issue the challenge again to anyone who reads this. I will debate anyone, anytime, anyplace, on the veracity of the theory of evolution and the existence of an Intelligent Being, and make them look like a fool. I am that confident of the utter absurdity of their vacuous position. I don't expect anyone to take up the challenge because I've found that evolutionists don't really want an honest debate, they only want to shut you up. They have already effectively censored creation out of the schools, and even the public arena for that matter. This, from people who profess to worship at the altar of free speech and academic freedom. They believe in these concepts only when you adhere to their viewpoint. This is why they're not really educators at all, because science follows the facts wherever they lead. If you don't like where the facts lead, and you censor all opposing viewpoints, then you are not a scientist, and not even a true liberal, let alone an educator.

I have practiced what I preach. When I was a teacher, I would allow my students to have an honest debate about evolution in my class. I divided the classroom into those that believed in creation and those who didn't. I had the two groups express their opinions one at a time without interruption. At no time did I interject myself into the debate, or verbally bludgeon any student who believed differently than me. My only role was to get the students to question whatever "facts" they thought they knew, and which they were using to draw their conclusions, and make available to them the overwhelming science that refutes such an unfounded theory. Once they had the facts, it was up to them to decide, as I was just a facilitator. It was a great exercise in academic freedom just the way education should be. I didn't ask permission from the PC educators to have the debate in the first place, or they would have tried to stop me, even though I was violating no law. I didn't ask permission because on two previous occasions that I had dismissed evolution as having absolutely no scientific basis in front of a classroom, I found myself being verbally reprimanded by one of our principals, even though I had done nothing legally wrong. Understand that these principals didn't necessarily disagree with my beliefs. They were simply intimidated by students who complained that I was violating the state-sponsored orthodoxy they had been programmed with since primary school, and they just didn't want any controversy. I remember telling one of the principals that I intended to continue teaching <u>science</u>, not religion or philosophy, and I was there to give kids the facts. I even started a Creation Science club at the school, where during our activity period, I could disseminate much more information about the subject, free of the restrictions I had to endure in a regular science class.

It doesn't stop with evolution. When you don't believe in God, you are left in the untenable position of believing that we, as human beings, are the product of billions of years of such an incredibly fortunate series of mutations, the collective odds of which would make the lottery look like a sure bet. This has moral as well as scientific implications, because it then follows that we're just an accidental gob of chemicals with no moral authority other than ourselves. After all, we are the highest life form, and the most intelligent beings on earth. And since we don't have any higher moral authority than ourselves, then what's to stop us from doing whatever is right in our own eyes, or whatever makes us feel good. That's why evolution, coupled with the moral relativism and values clarification teachings of affective education mentioned earlier, is a dangerous combination to instill in a young mind. If a student is "dissed" by their classmates, and wants to get even by shooting them, why not? If a student hurts others in their quest to satisfy themselves with drugs or recreational sex, why not? But, you might argue, even without a belief in a

Higher Authority, we all have to submit to, and respect, the legal authority of law in order to have a peaceable society where we can all live together (group-think). I would in turn argue that those laws all came from the moral code and Judeo-Christian worldview found in the Bible, which was ultimately codified into English common law and American law.[83] If kids are taught that there's no Higher Authority than themselves, then why should they obey laws that spring from any other Authority than themselves, let alone one that originated from sources beyond his or her experience? A person who rejects God, certainly won't feel constrained by any government. It is no coincidence that Darwinism had a great influence on the philosophies of Nazism, racism, eugenics, abortion, and euthanasia.[84]

I hope you can see that the moral implications of rejecting a Supreme Being are very serious indeed. I offer the history of the current generation as evidence. Since school prayer was banned in 1962, there has been a meteoric rise in the rates of violent crime, drug abuse, pre-marital sex, teen pregnancy, and sexually transmitted diseases in our schools and society. If you look at the statistics for these categories and plot them on a graph, for many years prior to 1962 the line was flat or declining. After 1962, the rates increased exponentially, 400-600% for each of these categories for the next 25 years with no signs of slowing down, even after factoring out population growth.[85] Coincidence? I don't think so. Yes, I know there were other factors, but there is no denying that it was A factor, and likely the single greatest factor. What other events during the sixties would cause a nation to lose its innocence and morality so drastically? The

Vietnam War? Probably a factor, but a war the public tried to ignore. President Kennedy's assassination? Hardly. Availability of birth control? For adults, yes, but birth control was not widely available to high school teens until 15-20 years later, and birth control had nothing to do with the increased rate of violent crime. Too much affluence? That may be a contributing factor, but not the root cause. Even the Kinsey Report, largely credited with the sexual revolution, was published 10-15 years earlier, and though it had an undeniable influence, the statistics I'm talking about don't even start to show an increase until the following decade. You can think what you want, but I prefer to believe that <u>personal morality</u>, which is what we're talking about here, was much more influenced by a series of court decisions from 1962-1972 on school prayer, birth control, pornography, abortion, and relaxation of drug laws. These "laws" enacted by unelected judges taught kids that they came from nothing and they are going back to nothing (leaving no logical person to believe life has any purpose). Kids learned that women are sexual objects, and that there are no consequences for promiscuous behavior. They learned that there is no value in life at all, since we're just a mass of chemicals anyway. With this kind of "education," why then should they act any other way but in a completely selfish manner? And we wonder why kids are killing kids, men are abusing drugs and women, and moms are killing their babies? During this time period our country chose to officially reject the authority of God in matters of public discourse. No living, breathing, red-blooded American can objectively conclude that our country has been better off since. In fact, every survey continues to show that the

American people overwhelmingly feel we are going in the wrong direction.[86]

We now have an entire generation of educators that have a predisposed hostility towards anything religious. This is all the more incredible considering our roots as a nation, and the avowed purposes for which most of our nation's schools and universities were founded.[87] Noah Webster, one of our nation's leading educators, said, "In my view, the Christian religion is the most important and one of the first things in which all children, under a free government ought to be instructed...No truth is more evident to my mind than that the Christian religion must be the basis of any government intended to secure the rights and privileges of a free people."[88] Benjamin Rush, a physician, and signer of the Declaration of Independence said, "The only foundation for a useful education in a republic is to be laid in religion. Without this there can be no virtue, and without virtue there can be no liberty."[89] I could go on and quote many of the other Founders, as I have done elsewhere in this book, who have said similar things about the relationship between religion, morality, and government. Even William O. Douglas, the most famous "liberal" Supreme Court Justice in history, said "We are a religious people whose institutions presuppose a Supreme Being."[90] Failure to acknowledge God in the public arena because of the Supreme Court- invented myth of "wall of separation of church and state," is not only an incorrect interpretation and application of law and history, it has had grave consequences for our nation. Incidentally, the phrase "wall of separation of church and state," is not found in the Constitution as so many people have been led to believe - a point made clear by the dissenting justice in the Supreme Court's 1962 case, **Engel v. Vitale** which struck down school prayer.[91] In fact he called the phrase a "metaphor" and its use by the Court irresponsible.[92] What the Constitution does prohibit is the government's establishment of a state-sponsored religion as well as their interference in one's free exercise of religion. Many of the Founders have made it abundantly clear that religion and morality are indispensable to the maintenance of a free republic.[93] If the First Amendment truly intended to ban all religious expression from government, including public schools, then why did the Founders' own well-documented actions and words show that they believed just the opposite? And despite this, we have had open, repeated hostility toward any religious expression in our schools in the form of banning certain student attire, to equal access to school facilities, to censorship of student's written and oral classroom work, to censorship of valedictorian speeches – events that have become too commonplace to even list them all. Typical is an incident in a Nevada school district where school officials actually turned off the microphone of a valedictorian in the middle of her speech, rather than let her share her belief in Jesus, while her classmates cheered, "Let her speak!"[94] Exactly what was it about crediting Jesus for her success that could have had such a corrupting influence on the student body, when these same students have heard that name used as a curse word their entire lives? How hypocritical of these school officials, who claim to want what's best for the students, and who supposedly value freedom of expression above all else!

Set Our Children Free

And there are other moral implications to leaving God out of the public arena. If we wholly subscribe to the Godless humanistic point of view, then we believe that nobody outside ourselves is in control of the world and universe. It then follows that if no one is in control, then the delicate balance of life here on earth that we enjoy is always in peril, because it just happened by chance anyway. That is why the chicken- little folks like Al Gore can have a following that would make the pied piper envious. These folks are so convinced that a disaster is always lurking somewhere in the cosmos waiting to wipe out all life on earth as we know it. In the sixties it was the chemical scares of DDT and chlorofluorocarbons. They managed to get those chemicals banned without a shred of evidence that they were harmful. The truth is that these chemicals were beneficial, and banning them actually cost millions of lives.[95] In the seventies came warnings of acid rain, ozone layer depletion, and a new ice age. Just ten years later, these theories had largely been debunked, and now global warming was the threat.[96] Understand that despite everything you may have heard, there is no evidence whatsoever that supports any of this junk science.[97] If anyone reading this doubts the veracity of what I just said, then I challenge you to do the research yourself and debate me if you like. I don't expect any takers, because unfortunately, the "green" crowd doesn't seem to want an honest debate either. Every science class I taught, I managed to find a little time to educate my kids on these issues, and guess what happened? About half of the students actually argued with me, and questioned that I knew what I was talking about! Think about that for a moment. I was the teacher, who had done the research on the subject, and yet some of my students who had done none of the research, were offended to the point of being angry at me for telling them the truth. Not one of them knew any of the underlying science, other than listening to the propaganda they had been force-fed since kindergarten. To them, I was a crackpot who wanted to rob them of the opportunity to save the planet someday and add some meaning to their lives. That's how pervasive PC has become in our schools.

And if the kids have been that easily swayed by the environmental agenda, do you think the sexual revolution has had less of an effect? It's bad enough that students see and hear sexual messages every day from their peers and the media. They are further inundated with sex education that includes much more than anatomy and physiology. Sex education in many school districts includes not only information about condom use and birth control, but the actual dispensing of birth control.[98] Some school districts will even transport girls to abortion clinics without their parents' knowledge or permission.[99] How did we get to the place where a student has to have a doctor's prescription to bring an aspirin to school, or a note from a parent to go on a field trip, but no permission to get an abortion? Schools feel free to not only teach kids about alternative lifestyles, and enforce speech codes against it, but routinely deny those same liberties to Christian students who are forced to silently accept beliefs that are anathema to their own. Even when kids want to wear Christian-themed T-shirts, meet on school grounds, or express their beliefs in any number of non-intrusive ways having nothing to do with the official

Set Our Children Free

curriculum, the first response of many knee-jerk school administrators is to deny them rights that are protected by federal statute and the U.S. Constitution. Not only are they mocked into silence, but prohibited by law from having their own viewpoint represented in the curriculum. Parents, don't assume that your school is any different. If your school is typical, you will have to ask a lot of tough questions of your school administrators to get the truth. Expect evasiveness and misinformation. Trust me, they're not going to tell you stuff that will stir up the community if they can help it. If your school is teaching "comprehensive" sex education, you're not going to find out what they're teaching unless you have someone on the inside (a teacher friend or your kid) that you can trust to tell you the truth.

And if that wasn't enough, we now have to put up with our children being exposed to the utopian myths of multiculturalism and diversity. Let me preface this discussion by first giving you some information about me personally. I am a first generation Italian American, whose father came to this country through Ellis Island – legally – and whose mother was born shortly after the arrival of her Italian father. My father and mother's family left Italy and embraced America and all she stood for. After they settled here, they didn't even teach their children the Italian language, because they knew that they had found a better home than the one they left. Ultimately, they considered themselves Americans first, and Italians second. They became part of the great melting pot that formed the backbone of our nation in the early twentieth century, even though they faced overt discrimination and a Great

Depression. And amid all the poverty and prejudice, none of them advocated solidarity and "social justice." They knew that America was a place where their hard work would someday be rewarded. Having said all that, I believe that whatever the well-meaning motive for teaching multiculturalism and diversity studies in our schools, the result has been divisive, not unifying. The foreign exchange students who attended Hancock wondered why we called people "Irish-American" or "African-American," etc. They didn't use such language for foreign citizens in their home country. There's nothing wrong with learning about one's heritage, but only in the context of the role that heritage can play in contributing to the vitality of our great nation. Too often such classes are taught about what a racist country this is, and how minorities are not being afforded social justice. These classes do not belong in high schools, which again should be devoted to academics. Even in college, these classes are frequently taught by social revolutionaries. The bottom line is that our national identity should trump any racial or ethnic identity, and teaching otherwise is factionalist and dangerous. Germany and France are now finding out the hard way that multiculturalism isn't working.[100] Other European countries just haven't gotten around to admitting it yet. Look at the grief Arizona has gotten over two laws that they just passed.[101] The first, which received a lot of press, was simply a law enforcing an existing federal statute. This brought ridiculous charges that the law encouraged racial profiling, and rallying cries to boycott the state. The Los Angeles Unified School District school board even went as far as passing a unanimous resolution condemning the law, and calling upon all students in their school district to be instructed that the law is un-American and racist.[102] They also condemned the second law, which received very little publicity, but bans schools from teaching classes that promote the overthrow of the government or advocate ethnic solidarity.[103] Did you ever think that you would live in an America where an <u>educational</u> board would not only openly advocate breaking the law, but advocate teaching our children to promote ethnic solidarity and the overthrow of our government? This is revolutionary stuff and shouldn't even be taught in college, let alone high school. Can you imagine the outrage if there were public schools teaching classes on white separatism, and the overthrow of our government? I don't need to tell you that those school administrators would be fired and probably prosecuted for teaching such conspiratorial trash in the schools. Similar groups have been prosecuted by the feds for doing the same thing on private property, let alone on public school property.[104] And before you pull out the race card on me, you need to know that I adopted two bi-racial children 30+ years ago, and I have a diverse group of grandchildren that would make the United Nations look like a close-knit fraternity by comparison. I love each and every one of them dearly, and if need be, I would kill for any of them.

And speaking of the United Nations, I believe they are the greatest threat to our liberty that exists today. It is not difficult to discern from reading credible news sources, that the pieces have now been put into place for a world government. Many countries have handed over their sovereignty to the U.N. by signing treaties on climate control, international justice, the rights of children, food distribution, and health, to name a few, and our country is no exception.[105] A treaty signed by the President and ratified by two thirds of the Senate has the same force of law as our Constitution. By the way, when I say "credible" news sources, I exclude all of the network news and almost all newspapers. None of these sources practice true journalism anymore, and in fact, are in the business of promoting their propaganda and censoring legitimate news at the behest of our corrupt government. The latest outrage from the U.N. is the International Baccalaureate Curriculum, which combines all of the elements of the PC curriculum I have described above into a program that has already made its way into many American schools. Couple that with the Race to the Top (RTTT) program funded by the so-called "Stimulus" bill passed by Congress in 2009, and you have what amounts to not only a complete federal takeover of our schools' curriculum and every aspect of teaching it, but a de facto international takeover.[106] The curriculum indoctrinates kids with the same old socialist, humanist, globalist, radical environmentalist, multi-culturalist nonsense that the schools now teach piecemeal, into one re-packaged piece of garbage under a new name.[107] These programs, and the federal control that comes with them must be purged from our schools, and their mandates resisted at all costs. Most people aren't even aware of the magnitude and breadth of this educational revolution since it has been done quietly and with the complicity of the leftist news media, which is supposed to be a government watchdog. The only fuss the media made was when one of the states – Texas – dared resist and write their own standards.[108] There was an uproar about how Texas was trying to change their textbooks, and nary a peep about how the Obama administration was co-opting the rest of the nation's schools. Does this scare you? It should. Thomas Jefferson wisely noted, "When government fears the people, you have liberty, but when the people fear the government, you have tyranny."[109] Taking into account all that I have discussed above, the actions of our government can properly be described as nothing less than tyranny.

Chapter 9

HOW WE CAN FIX IT

The solutions I propose are radical because half-baked reforms will not cure our nation's schools. Desperate times call for desperate measures. Nothing can be gained from listening to people who are part of the problem, and by that I mean the educational establishment. As an engineer, I have been trained to find the most efficient solution to a problem, not the most popular. Compromises are popular, but the answers to our educational dilemma won't be, at least to those with a vested interest in the current system. And if we don't strive for the best solutions, we will accomplish nothing at all. Our educational system - check that - our <u>nation</u> will be radically changed for the better if we have the courage to do what I am proposing. I am a realist, yet I can envision a day when public schools will be free from the shackles that have bound our children from getting the education they deserve. Before I discuss my seven point plan, I want to give you an idea of what our schools can and should be, starting with the first day of school. When the chains of federal, state, and local regulations as well as the monopolies of the teachers' unions are broken, the schools should look something like the one I describe below on opening day. On the first day of school, there would be a mandatory assembly for all students and their parents. The first and only order of business would be the following speech given by the local school administrator.

Set Our Children Free

What the First Day of School Should Be

"Thank you for making the decision to attend our school this year. We also want to thank all of the parents who have taken the time to attend. As you know, the state legislature now permits you to spend your tax money to send your kids to any school in the state. There are a variety of public, charter, and private schools you could have chosen. We thank you for choosing us."

"As a student here, you will be required to pursue one of several career paths: college, vocational, or business. It is our expectation that this will be your primary focus while you are here. This focus must be reflected in your attitude, your behavior, and most of all, your willingness to work diligently towards your career goal and the intermediate milestones that our school has established. There are no K-12 grades here, only state-mandated academic milestones. Once you have completed all of them, you will have earned your diploma and be free to pursue a career or apply to an institution of higher learning. If you fall behind on the timelines for completion of your milestones, or drop below the academic performance standards we have established, you will be given one semester to remedy the situation. If you fail to do so, you will be asked to find another school that is better suited for your educational ambitions or lack thereof. We demand that our teachers set high standards, and back this up by assigning you enough material, and drilling you repeatedly in that material, until you have the requisite knowledge and skills to meet these goals. Our teachers will also assign several books of reading material for you to complete during the summer recess as preparation for the following Fall semester. We will assess whether you have met your milestones by means of testing you as an individual, never as a group. The teacher is the final authority on your grade, and since they are in the best position to observe and evaluate your progress, we will never overrule them."

"Our curriculum is second to none in preparing you for the next stage of your life. Those students who have tested below our established reading and writing standards will be put into an intensive program, and not permitted to take any other classes until their reading and writing skills are up to par. It is expected that by the time you receive your diploma, you will read, write, and speak the English language in a manner that will not only prepare you for the job or college of your choice, but reflect pride in this institution and this country. And although you will have the opportunity to learn several foreign languages, you will only learn one culture – what it means to be an American. Any other separate racial or ethnic identity you possess is only relevant in regards to the gifts, vitality, and strength that culture can contribute to the national unity and pride all of us should experience as Americans."

"Understand that our curriculum is fact oriented – facts which will require extensive reading and memorization on your part. There are no affective education programs here, only cognitive classes, in which we will impart knowledge, not practice psychology. The math curriculum is not a one-size-fits-all approach, but is tailored for each career path –

college, business, or vocational. Students in these different career paths will be taught only the type of math applicable to that chosen field. Science studies will be devoted strictly to proven factual information and the search for new information. Theories will only be discussed in the context of developing the discovery thought process within the framework of the foundational knowledge that students have already been taught. While we practice academic freedom, and therefore no theory is off limits for discussion, we will not teach unproven theories as fact. Our history textbooks have been written from factual data derived only from primary sources, and have been purged of political content and comment. There will be heavy emphasis on our nation's exceptional heritage, our rights and responsibilities as a free people, and the importance of our founding documents. In economics you will learn the principles that have engendered the highest standard of living ever known. Physical education classes will be same-sex, and will focus on the physical and physiological development of the body. Sex education will focus on the responsible moral, medical, psychological, and sociological reasons for remaining abstinent until individuals are mature enough to marry and raise a family."

"All of our school rules exist only for the purpose of facilitating the best possible learning environment. In keeping with that purpose, none of our classes are co-ed. In addition, male students are taught and coached only by male teachers, and female students are only taught and coached by female teachers. Our experience has shown that this arrangement is much more conducive to learning. Also, our learning environment does not permit cell phones, cameras, i-pods, or similar devices on the student's person until the end of the school day. Similarly, our dress code will be strictly enforced, since violations of the code distract from the purpose of this school. Students, you will have five minutes between classes, twenty minutes for lunch, and one study period. There will be no wasted time. You are not here to pursue a girlfriend or boyfriend, catch the eye of college scouts, or campaign for homecoming queen. These things may indeed happen, but they are incidental to the primary purpose for which you are here. And to further enhance the learning environment, bullying will not be tolerated. Any reported bullying or any altercation will be thoroughly investigated, and the instigators will be expelled permanently. Those students acting in self-defense will face no consequences."

"In addition to being trained in the skills you will need in order to be successful in college, business, or a vocation, you will be taught how to conduct yourself in a manner that will earn you the respect of recruiters, employers, and other productive members of society. Your behavior, attitude, and demeanor while you are here, whether inside or outside the classroom, will be exemplary in adherence to both the letter and spirit of the school rules. Three documented violations of the school rules will result in expulsion. There will be a brief probationary period during which you will be allowed time to acclimate yourself to our rules. After that, any teacher or administrator will be the final authority as to whether your actions are rules violations. They will be encouraged to give a verbal warning first, but once a violation is documented, it is final. Any arguing or disrespect towards any teacher or administrator is a rule violation, and if the disrespect is serious enough, it will result in immediate expulsion for the first offense. If a student commits a criminal act, they will be expelled immediately. Our teachers and administrators have the full support of the school board and state law. Their instructions are to be followed without question unless they ask you to do something illegal or immoral, in which case it should be reported directly to the principal. Conversely, if you make a false accusation against a teacher or staff member, you will be expelled permanently. Bad behavior will not be tolerated as it is unfair to the majority of our students here who are serious about getting an education. Our teachers and staff are here to help you, and they do so in many instances at great personal sacrifice. Therefore, it is our expectation is that you give them your complete respect and cooperation. As you may be aware, there is now no state compulsory education past age 14. The law also does not require any school, other than our alternative school, to accept a student who has been expelled. Therefore, if you are dismissed, your only options may be home schooling, finding a job, or trying to survive the alternative school."

"Parents, by enrolling your child here, you agree to support all of the school's goals and rules. You are our partners in this endeavor, and we will do everything in our power to earn your trust. Besides this meeting, you are expected to meet with your child's teacher at least once per semester at the designated day or evening sessions. Failure to do so can result in denial of your student's application to return the following year. We are open to your suggestions about how to make our institution better, or to help your student maximize their individual talents. What we will not do is listen to complaints in the context of mitigating some academic or behavioral shortfall on the part of your student after-the-fact. By enrolling your child here, you agree that at the high school level, they will assume full responsibility for their education. At the primary school level, teachers will assume the majority of responsibility, with a decreasing amount in each succeeding year until high school. Because we expect much from our students, we expect even more from our teachers. We offer a very competitive wage in order to attract the best educators, and we offer higher incentive pay for the most skilled among them. In return, they are continuously evaluated for both their instructional skills as well as their behavior, by direct or remote classroom observation at least once per week. In other words, we are continuously in the process of monitoring and improving the quality and professionalism of our instructional staff. Our contracts with our teachers are year to year, with no tenure. If they do not meet our high standards they will not be asked to return. If they do meet them, they will be rewarded accordingly."

"In summary, we are a professional educational institution, and we expect as much from ourselves as we do from you – the parents and students. In return for your money, trust, and high hopes, we will do our part to give your child the best possible preparatory education available. Our reputation is such that if your child is successful here, there will either be a job or a college waiting for them as a reward for their hard work, since they will be thoroughly prepared for this next phase of their lives. A diploma and referral from our school is the best thing they can have on their resume'. Thank you for your support and let's have a great year."

I guess you can tell from the above, that the changes I am advocating will indeed be considered radical to the educational establishment, although they may sound more like common sense to some of you. I make no apologies because tough medicine is sometimes needed to restore health. Here are the reforms I propose along with my reasons. The last reform, #7, is the most important, and offers the blueprint for enacting all of the other reforms. You may not agree, but you have my open invitation to debate me on these issues at any time.

1. Empower Taxpaying Parents - Let me see if I got this straight. The state takes our hard earned money and gives us no choice as to how we want it spent on our own kids' education? How did we allow this to happen? It's our money and our kids! Give us back our money and let us decide how we want them educated and where. Government is supposed to be our servant, not our master. The only interest the state or federal government should have in education is to require us to provide one that meets certain minimum standards. Why should the government have a monopoly on education? The U.S. Constitution gives the federal government no role in education, and some state constitutions are limited to providing a "free" education, not a monopoly. And since we are paying for it, why should we be taxed to death to support a broken system that doesn't produce competitive results, and indoctrinates our children with political and moral lessons that are objectionable to most of us? Enough! We need to demand an end to property taxes to fund schools, a system which pits property owners against teachers, firemen, and police. Let parents keep their portion of school taxes (stop calling it a "voucher," it's our money) and pay their own tuition, or simply fund schools with a sales tax instead of property taxes. Sales taxes would still be "progressive" since the rich spend more money and thus will pay more taxes anyway.

Better yet, we need to simply demand tax credits for those parents who choose to send their children to private schools. Using the recent **Arizona Christian School** Supreme Court decision mentioned in Chapter 6 as the legal basis, we need to apply pressure to the many newly-formed (2011) state legislatures to do just that. It is the perfect vehicle for empowering parents to choose the best schools for their children. And why should parents get a tax credit? Because every child removed from public schools is one that the government doesn't have to spend money to educate. The Supreme Court in the above case also made that same point. All schools – public, private, or charter – should be funded on the same per-pupil basis through tax credits or direct appropriation. Parents would choose one of these schools for their kids each year. To those of you who would argue that you don't want your tax money going to support a religious viewpoint with which you disagree, I would counter that many parents are required to do exactly that now by having to send their kids to the nearest public indoctrination center. The difference is that you don't have to send your

Set Our Children Free

kids to a private religious school, but many parents presently have no choice but to send their kids to public schools, as they can't afford anything else. The news media and the teachers' unions love to characterize the school choice debate as using "public money" (read - your taxes) to fund private religious schools, and trot out polls that show a majority of people would be opposed to this. Word the question differently, (i.e. Would you favor using <u>your own</u> tax money to send <u>your</u> kids to the school of <u>your choice</u>?) and you would get the opposite result. And despite what activist judges have ruled, there is nothing unconstitutional about this. In fact, using our tax money to support <u>only</u> public schools is discriminatory against religious schools, which makes this public policy in itself unconstitutional. The recent **Arizona Christian School** Supreme Court decision has finally resolved that question.

All schools, including public schools, should be free to compete for students and teachers. Existing school facilities should be turned over rent-free to those schools who signed up the most students prior to the start of the school year. Alternatively, the laws in many states now allow existing schools to be turned into charter schools if the community supports it. This is a golden opportunity to have a taxpayer funded school without the burden of many of the government regulations. In either case, charter school laws must be strengthened so that they not only receive the same per-pupil funding, but that they have the same access to public school facilities and public school transportation. And teachers who switch to charter schools should be allowed to stay in the state retirement system, and not penalized.

Set Our Children Free

These are public schools after all, funded by the taxpayer. Schools would then have the freedom to attract and pay the best teachers at any salary they choose, and all teachers would be on year-to-year contracts. State lawmakers and school boards need to do their part by making it easier to fire bad teachers, and by all means quit helping the unions out by providing automatic deductions of dues from their paychecks. Instead of a regional school board, schools should be locally run and managed, answer to the parents of each school, and thus be more responsive to any parental concerns. If a parent's objections aren't heard, they can take their tax dollars with them to another school. If enough parents move, the school administrators and the school will no longer be in business the following year. Any non-academic curriculum should require approval by a majority of the parents whose children attend that school. If you think this is far-fetched, re-read the examples I give in Chapter 6 and realize that the current trend is favoring exactly the kind of school choice I'm talking about, both in this country and abroad.

Teachers' unions have long argued that vouchers, school choice, or the like, would mean the end of public schools. Nonsense. We would still have the same number of teachers and the same number of students. They would just be reassembled into different buildings and classrooms, free of the burdensome regulation and union monopoly that is currently stifling education. Each of these schools would be free to innovate in any way the marketplace decides. There may be an all-boys school or an all-girls school for instance. There may be schools that cater strictly to college-bound,

Set Our Children Free

and some that exist only to prepare students for a vocational trade. There would be some schools that focus on a particular area of art or science. They would be free to require uniforms, shorter class periods, and any curriculum that meets state milestones. Schools would be required to teach to certain academic standards in each subject, and the students would be tested to assure those standards are met. This would not change from what we have now, except that the standards need to be much more rigorous, not some minimum requirement that most kids can pass in their sleep. A diploma should mean something. And forget about K-12 grades. Milestones should be set so that when a student has learned the necessary material required by the standards, they could complete that milestone by testing and move on. Those students that achieve certain milestones by age 14, for instance, can choose the college curriculum, and those who don't can pursue the business or vocational curriculum. When students complete all of the milestones, they receive a diploma, regardless of age. This rewards the ambitious students, who can go as fast as they want. Aside from requiring these standards and milestones, the government would be out of the business of education entirely.

Some have recommended that parents pull their kids out of public schools and let the weight of the educational bureaucracy collapse the system, but I'm not so sure it would work. Bureaucrats will find a way to spend our money even if there are fewer students to spend it on. Not that I think it's a bad idea for parents to abandon the public schools, but I think there's a better chance for even more substantial changes if an informed electorate demands it. The recent

Tea Party movement is a perfect example of how ordinary citizens can effect change. In fact, the Tea Party movement, along with other worthy grass roots organizations, might be the perfect vehicle to empower parents to make the needed changes to our schools. After all, the school reforms that I advocate synchronize nicely with these organizations' values of fiscal responsibility and a return to Constitutional principles. Regardless of the vehicle, however, it is absolutely necessary to organize like-minded parents, educate others on the issues, demand change, and hold elected officials accountable for making these changes. I don't pretend that it will be easy, but the alternative is to grow old in America seeing your children and grandchildren enslaved.

2. Empower Teachers - Before you jump to conclusions, I am not endorsing any of the agenda of the national teachers' unions. However, the problem with many of the past attempts at school reform is that lawmakers don't listen to many of those who know what needs to be done – individual teachers, not the teachers' unions. The major unions have their own radical leftist agenda, which is why they have lost all credibility with lawmakers. We must understand, however, that the unions do not represent the views of the average rank and file teacher. Many of them have joined the unions for job protection, not because they necessarily agree with them. And the good teachers are just as anxious to get rid of the bad teachers as the rest of us. Granted many teachers lean left in their politics, especially those in the unions. But what I am proposing below, I guarantee, would be supported by the vast majority of instructors who have to face those daily battles in the trenches.

By empowering teachers, I mean that we should give them largely unfettered freedom in two areas: 1) the ability to fairly assess the students academically, and 2) control of the classroom. Make them the ultimate authority for a student's grade and behavior. Standards should be established by the state, but leave the teaching to those who know how to do it. And by all means, leave all discipline to them. As detailed earlier in this book, students have no reason to respect teachers, and absent that respect, the student has no reason to pay attention or behave. If they refuse to make an effort, teachers are expected to somehow compensate by lowering their standards and passing them. If students misbehave, they go unpunished and detract from the learning experience of others. It is because teachers have been neutered. They have no real authority and students know it. A generation ago, we feared teachers. There was no crying in education. If you got a bad grade because you were lazy, there was no ganging up on the teacher for setting the standards too high. And if you misbehaved and talked disrespectfully to a teacher, you got paddled and maybe expelled.

I wouldn't be in favor of bringing paddling back unless it was fully supported by state law. Not that I don't think it would work, but the last thing the schools need is a flood of lawsuits. If a teacher is given real authority, paddling wouldn't be necessary. By real authority, I mean that a teacher must be permitted to remove a student from their classroom permanently for anything from disrespect to law breaking, and if they do, it shouldn't be held against them. Our state gave teachers that authority, but most teachers dared not use it due to fear of retaliation from school administrators. The law should give teachers, and teachers alone, the authority for discipline, and the respect will return. The law also needs to prohibit any retaliation against a teacher for implementing any reasonable punishment for student misbehavior.

Real authority also means that the teacher has the final say on grades. Every principal I knew always insisted that we set high standards, but it invariably turned out to be lip service. When parents and students banded together because the teacher, in their eyes, was making them work too hard, the principal would always cave. Granted, teachers are human and they will be wrong on occasion. That's why we have a continuous evaluation process by the school administration. If a loose cannon gets too far out of whack, he/she can be reined in, but the admonishment should always be done in private. The public stance to the parents and students should always be: "Do what the teacher requires!" I'm convinced that 99.9% of all teachers, regardless of their political persuasion, are not in this profession to throw their weight around and bully kids. There just aren't that many head cases out there. All of the teachers I've ever met just want to help the kids, and they only differ in their methods. Give them a little rope. Besides, if proposal #1 above is implemented, we can get rid of the bad teachers, which will foster much more trust in the rest of them.

Set Our Children Free

And why do we require all teachers to have a baccalaureate degree? Many subjects can be taught by teachers with associate degrees and even, in some cases, a high school diploma with some additional training. We don't require a baccalaureate degree for substitute teachers. If subs are unqualified then why do we let them in the classroom for weeks and months at a time when there is no regular teacher available? The dirty little secret is that it doesn't require a 4-year degree to teach, any more than a degree is required for many other professions. That's simply what the certification holders (colleges) have convinced lawmakers to enact because it is in their own interest. There would be no teacher shortages if we would allow qualified non-degreed people to teach, as well as recruiting teachers from other professions. We also need to ditch the salary scale for teachers. Offer more money for teachers in critical shortage areas, like math, science, and special education, and you will have all the teachers you need. Some states already allow local boards to negotiate this directly with the teachers' unions. States should go further and simply allow teacher pay to be a free market decision by each school board and not a negotiated issue. And forget the incentive pay some states offer teachers for completing college credits or some other prescribed mandate that has nothing to do with their effectiveness as an instructor in the classroom – it is wasted money. Every time there is a school reform movement, there is a cry to upgrade the teaching profession by requiring more training, education, or other mandates from an already stressed out instructional staff – at their own time and expense of course. Forget that! Any teacher who has spent a year in the classroom already knows how to teach, just tell them **what** to teach and give them the authority to do it without interference. Local school officials are in the best position to know their teachers, so for better or worse, give them the power to reward their best teachers and get rid of the bad teachers – not on a whim, but for just cause. Of course this also requires the appointment of no-nonsense principals that will require academic and behavioral discipline on the part of both teachers and students. Demand this kind of accountability on the part of state lawmakers, school boards, and school administrators. If you can't articulate what needs to be done, give them a copy of this book and make them explain to you why these reforms wouldn't work.

3. Stop treating all kids alike – We somehow have developed this notion that every student has to go to college in order to reach their full potential and help us compete with the world. I think it's rooted in the liberal concept that all children are equal in every way, and to admit otherwise would be discriminatory. In their view, parents are the ones who mess up their children, and if only the state could take them at birth and raise them in a politically correct atmosphere, they would all thrive equally. This is pure baloney, of course. Whether you believe that intelligence or environment has more of an effect on a child's learning ability, the truth is that each child has a different gift. One size does not fit all, and some kids simply aren't suited for college and never will be. There are many types of intelligence, and some kids who are gifted with great mechanical, artistic, or social ability, simply don't have the attention span, temperament, or mental capacity for book work. They will be the small business owners, craftsmen, and salesmen of the future. Why burden them with a college preparatory curriculum and doom them to academic failure? We don't think this way about special education students. We have different standards for them, so why do we treat all other students like they're equal? Kids that just aren't going to cut it academically, for whatever reason, can be identified as early as middle school and put on a business or vocational track. Let them become proficient in the trades so they can earn a decent wage when they graduate, or put them into a business track where they learn computer skills and how to run a business. College bound students could then take academically vigorous classes, in which teachers would be encouraged to challenge the students, without having to worry about flunking half the class. For example, we currently require all students, regardless of their ability or interest to take Algebra and Geometry. Why? Why not just business math? If choosing a college, vocational, or business course upon entering high school was good enough a generation ago, why isn't it good enough today? We need to insist on this graded approach to education, so that all kids can reach the maximum potential that is right for them.

Also, there's nothing magical about 12 school years. Some kids can complete high school easily by the time they're 16 years old. Let them do it and get on with their lives. There is an unspoken fear that such a teenager won't be ready for college or a job at age 16. I disagree. A teen ambitious enough to complete all of the state requirements by age 16 is exactly the kind of teen you could trust to be a model employee, college student, or even an assistant high school instructor. Besides, if the college, business, and vocational career paths idea above is implemented, schools would be free to design curriculums rigorous enough, that very few students would actually finish by age 16 anyway. At least in recent years, many students have been permitted to take college credits while they are in high school, allowing a student who is ambitious enough to save the time and expense of their first year or two of college. I would take this a step further and allow that same student the ability to finish high school a year or two early if they complete all of their milestones.

4. Be honest with the curriculum – Stop the political correctness. Lawmakers must require schools to spend a good portion of their history curriculum on the founding documents like the Declaration of Independence, the Constitution, and the Federalist Papers. Teach kids how unique and special America is, and how exceptional is our long tenure as a free society. After all, the proper teaching of history is how we pass our values down from generation to generation. Emphasize the rights, liberties, and responsibilities of our free society by drilling students in the concepts found in our founding documents. We cannot have unity in our nation if we continue to celebrate multiculturalism at the expense of educationally assimilating young people into our great culture and history - a history, I might add, that every American should be proud of. Lawmakers and textbook committees must assure that their books accurately reflect our great and noble history. It isn't about politics, it's about the truth. School boards should refuse to buy textbooks that do not depict America in an honest light. There are many good alternative textbooks that accurately depict the truth about American history and get their facts from primary sources. As noted in an earlier chapter, today's school curriculum will be tomorrow's government policy, so we must turn this around now!

Similarly, students should be taught that our economic system is second to none and why. Only a free market system, liberated from the shackles of central government planning, with a fertile ground where entrepreneurship can flourish, can produce the standard of living we enjoy here. Students should be taught that we live in a great and exceptional country where freedom and opportunity reign. If you are a business owner, volunteer to speak to school kids about the role private businesses and government should play in our economy. A revolution is taking place right under our noses, and we must join the fight! Remember, we have entrusted our money and our kids to these educators and we should demand accountability.

Be honest with the science curriculum too. We are such hypocrites when we teach that the scientific method is the basis for discovery of the laws of science, and then turn around and indoctrinate kids with theories that completely ignore that method. It's bad enough that our children are being force-fed a diet of evolution and environmental junk science, but the product of this propaganda will inevitably be a God-less, government-run dictatorship that will rob us of our freedom and control every aspect of our lives. The problem is, we're not even allowed to have honest debates about these subjects. What we have instead is censorship, starting with the school districts and ultimately supported by the courts. I do not use the word censorship lightly because that's what it's called when a governmental entity won't even allow the freedom of an honest and open debate. What do you think would happen if we tried to muzzle teachers from, not just teaching, but <u>even debating</u> global warming or the theory of evolution? Educators and lawyers would have brain hemorrhages rushing to the courthouse door crying censorship and lack of academic freedom. Yet that is exactly what they have done to us. When lawmakers in a number of states have attempted to legislate merely an open, balanced view of these issues, they have been rebuffed by the courts.[110] This is nothing less than the 21st century version of the thought police. Such judges should be impeached, and the politicians who appointed them removed. A better solution is one which is currently in place in over half the states. These states allow merit retention votes on judges after they have served one appointed term. This system should be extended to the federal level, including Supreme Court justices. If judges insist on making law instead of interpreting law, they shouldn't be given lifetime appointments, and we should be able to vote them out.

5. Create a fertile environment for learning – I genuinely believe the reason most middle and high school kids have no interest in learning is because they've never been required to learn. Create the proper environment, and that will change. The proper environment means an atmosphere in which they know that they alone will be held responsible for their academic and personal behavior. No excuses. The tenor of

Set Our Children Free

every school should speak loudly in word and deed from the moment a student sets foot on its campus, that education is serious business, and that laziness, foolishness, and misbehavior will not be tolerated. And that philosophy must be supported by parents, teachers, and school administrators. This can only be possible if points #1 and #2 above are implemented along with this one. Only when schools are free of government regulation can they set rigid standards and goals for students to meet, and jettison the ones who aren't serious about meeting them. Schools should not be babysitting institutions. If kids have reached the ninth grade and only come to school to interact with their peers, then they don't belong there. Behavior transgressions should be cumulative, like points on a driver's license. If he/she is guilty of three behavior-related offenses, they should be dismissed and put into an alternative school that is run like a boot camp. If they can't handle that, then let their parents decide whether to send them to work or home school them, which is easier than ever with online schools now. These are the same kids that cause the most disruptions, so ridding the school of them would create a much more fertile environment for learning. Under this plan, the derelicts would be gone and teachers would be free to challenge students and not worry about dumbing down the class material so that everyone passes. The rest of the kids would welcome the opportunity to be a part of a group where everyone has the same goal, without the distractions that the usual suspects create.

A proper environment also means one where bullying, drugs, and troublemaking will not be tolerated. Administrators seem perplexed about how to deal with bullying, but it's really not that complicated. A little hard-nosed investigation into each reported incident is all it takes. In most cases, administrators know who the bullies are, and they need the freedom to get rid of them. Make an example of the first few violators and the rest will get the message. School grounds are special places under the law, where a teacher doesn't have to have probable cause to search a student, and you don't need to prove something beyond a reasonable doubt. Use that authority, and furthermore demand that every school have the right to remove any student permanently who disrupts the educational process in any way. If a student is threatened and there are witnesses to that fact, the bully should be warned that they are on probation from that moment on. Any further threats or incidents, they should understand, will result in permanent expulsion from that school, and that assaults will be vigorously prosecuted by law enforcement. Even if there is a dispute about who started what, school administrators can find out the truth if they really make an effort, instead of just punishing all of the actors involved. Most kids will tell the truth about a bullying incident, especially if guaranteed anonymity, because they want the bullies gone also. Whatever the approach, the emphasis should be on dealing with the bully, not mandatory sensitivity training for everyone else like many schools advocate.

Without bullying you've eliminated much of the reason many kids resort to gun violence as retaliation. Unfortunately, for whatever reason a kid decides to bring a gun to school, he knows that a gun-free zone awaits him, where none of his

Set Our Children Free

victims will be armed. Notice that in every school shooting, from Columbine to Virginia Tech, the law prohibiting possession of a gun on school property didn't stop the law breaker. All the law did was assure that the shooters met no resistance, because law abiding kids and teachers obeyed the law. Some paid for this obedience with their lives. One armed teacher with a concealed carry permit could have saved some of them. By banning guns from campus, we have inherently made schools unsafe for anyone intending harm to our children. We walk and mingle among many concealed carry holders every day and we don't feel threatened, so why are we afraid of allowing responsible people, properly trained, to be armed so that they can protect our kids? In Israel and Thailand, where teachers are armed and trained, school shootings are virtually non-existent, including those from terror attacks. [111] An additional question to those who say that arming teachers is a bad idea – If an angry kid with a gun shows up at your child's classroom door, would you rather have their teacher armed or unarmed?

With the bad actors out of the way, schools would be free to really challenge students and have them start taking their education seriously again. Kids would have to study to make the grade or find another less rigorous education alternative. Schools should even require students to read several books over the summer vacation in preparation for the Fall semester. Again, the purpose of creating this type of environment is not to make things difficult for the student, but to assure that they will learn. The same can be said for the dress code. If school officials are not courageous enough to enforce a dress code, then require kids to wear uniforms – modest uniforms. Kids and teachers simply don't need the sexual distractions that exist on our campuses today, as it totally detracts from the learning environment. And there would be far fewer teacher-student affairs if we only had male teachers for male students, and female teachers for female students. Our schools are supposed to be educational institutions, not singles clubs.

6. Stick with academics, not social work – A typical home schooled child can learn as much in two hours as a public school child learns all day. This is fact, not conjecture. I've taught in a public school, a private school, and a home school and I stand by this statement. In addition to the personal attention a home schooler gets, they don't have to have their education diluted with all of the non-academic distractions and social engineering that goes on in public schools. It's also one of the reasons home schooled kids on average are several grades ahead of their public school peers, and perform so well academically that they are coveted by college recruiters. Parents, don't downplay the influence that an authority like the public schools has on your kids. At the very least, they will question, if not downright reject, everything you've taught them if it conflicts with what they are taught at school. And what they're being taught is that racial, sexual, and cultural diversity is always good to the point of being sacrosanct, in and of itself. "Diverse" individuals can never be bad, and if you think so, you are a bigot or racist. Your kids will likewise be indoctrinated into thinking that socialism and "social justice," the enforcement arm of socialism, is similarly on the protected species list of political philosophies. Even when these ideas aren't in the curriculum, they are modeled and peer pressured into the students in insidious subtle ways. Sadly, public schools have ceased to be academic institutions, and more often resemble social engineering gulags. Parents should be angry, but they only get that way when some outrageous incident makes the news – the type of incident that gives them a sneak peek into what's really going on inside those walls. School officials then downplay the incident to convince everyone that what happened was the exception rather than the rule. Parents, believe me, it's the rule. That's why I wrote this book. Nobody really knows what goes on in the classrooms except those of us who worked there. And trust me, political correctness rules and is rampant. Even teachers succumb to the pressure. I've sat in professional training seminars conducted by highly paid educational consultants, where I knew that 80% of the teachers disagreed with what they were being told, but most were too intimidated to speak up.

Review everything your child brings home, and also talk with them about everything their teacher said that they can remember. Pay attention to any announcement that the school sends home in a flyer – a means of communication that school officials love to use because they know that most parents will never read it, even if it makes it home. If possible, find a like-minded teacher or administrator who works at the school and ask them some tough questions. School officials love to implement programs or curricula on an opt-out basis, which requires no parental permission (and thus disclosure), instead of an opt-in basis, which does. When we fought to remove affective education programs from the schools, the one thing school administrators would never compromise on was giving parents the right to opt-in for their child. Their excuse was always that it was too much trouble to get permission from the parents, even though they have no trouble getting permission for far less important activities like field trips. They would rather have walked barefoot over broken glass than give back that kind of control to the parents. This is done on purpose because they know some programs and curricula are so controversial that you would object. Then even if your child opts-out, they are made to feel ostracized because they may be the only one. This is one of the many reasons you should never send your child to a public school unless there is no alternative. They may be spending seven hours a day learning things that completely contradict everything you've taught them at home, without either your permission or knowledge. That's in addition to the diversity and multiculturalism propaganda that, at best, divides instead of unifying students, and at worst, engenders revolutionary ideas that are dangerous to the safety of all students as well as our society. Until schools get the message that their first and only job is academics, they don't deserve your support. Re-read the last few paragraphs of Chapter 5 on the teaching profession. Every teacher knows that soliloquy is true. If that's what you want your teachers to be doing instead of drilling your kids on the multiplication tables or sentence structure, then you don't want your kids truly educated and you don't need this book.

7. Return to personal and Constitutional morality – Although this is listed as the last point in my seven point plan, it is far from the least important. In fact, this point is the most important of the seven. The first six points, if enacted, will result in cataclysmic change of our educational system. However, without a return to personal and Constitutional morality, these reforms will be short-lived. The reason is simple. What's wrong with our schools is what's wrong with America, and we can't change one without changing the other. If we could overturn the laws tomorrow, and force-feed these solutions upon the educational establishment, they would simply lie and demagogue their way back into power. All laws have their roots in moral principles, and right now our morality is out of whack. Despite what secularists would have you believe, much of the moral code that existed in this country came from the Bible, which eventually became incorporated into English common law, thus forming the basis of all of our laws.[112] Most of the schools in this country, as well as many of the prestigious Ivy League universities were founded for the sole purpose of educating the populace in the Scriptures.[113] If you read the Declaration of Independence and the U.S. Constitution, you will find that its principles of personal responsibility, self- government, and liberty cannot be separated from the moral and spiritual principles of the Bible. During the latter half of the 20th century, a prosperous America abandoned those principles in their personal lives, and we have been paying the price ever since. This decline in morality was precipitated not so coincidently with the advent of activist judges and a discarding of our Constitutional principles. The fruits of those disastrous rulings by unelected judges, have been irresponsible behavior and lack of personal responsibility and accountability, leading to sexual promiscuity, fatherless children, divorce, poverty, and crime.

It has dramatically changed our society within one generation, and by default, changed our schools. So now instead of teaching academics, we are re-educating a new generation to accept this "new norm" in the name of tolerance and compassion. And instead of teaching the value of national unity, the Pledge of Allegiance, and one nation under God indivisible, we are celebrating diversity and group identity. Instead of school prayers each morning reverencing "Our Father in heaven," students view *An Inconvenient Truth* reverencing our Mother Earth.

I made the point earlier in the book that most of the top performing kids come from stable homes where personal responsibility is not only taught but practiced. Conversely, nearly all of the bottom tier kids come from broken homes where a lack of personal responsibility and accountability is the rule. This is the generational pattern of immorality that continues to infect the school system. The other solutions I've proposed will work on the surface, and work for awhile, but there can be no lasting change to what ails our educational system until we deal with the corruption that our society brings into it from the outside. We have allowed this evil to prevail, to paraphrase Edmund Burke, because good men did nothing.[114] Each of us must now look in the mirror, commit ourselves to do the right thing on a daily basis, and pray for our schools and nation. Purging the corruption of our society up to the highest levels of government will require the same Divine help our Founders called upon when they declared their freedom. There are a number of good grass-roots citizen organizations committed to bringing about a return to sound Constitutional and moral principles in our country. Join one of them and urge like-minded friends to do the same. There is power in numbers, and once the numbers get large enough, elected officials will start to pay attention. The revolution has already started, as evidenced by the mid-term 2010 elections, and the resulting reforms in statehouses like Wisconsin, Ohio, and Florida, among others.[115] Many of those elected officials from both parties who have ignored this wave of freedom have already paid the price by being removed from power. But it is only the beginning of the culture war. The elitists will not surrender power easily or even peaceably in some cases. As these grass roots groups educate the populace, vet political candidates on their positions, and hold them accountable, change will occur. We are a self- governing nation whose power ultimately rests in the people. If a candidate for state or federal office will not commit to being faithful to the Constitution and assure that their appointed judges do the same, then find another candidate or run for office yourself. Have the courage to voice your opinion loudly and frequently to those already in elected positions. Trust me, they count the number of calls, letters, and e-mails they get, and extrapolate them into votes. The time for timidity is past. We can't make any meaningful inroads until we return to the principles of personal responsibility, self- government, and liberty inherent in the Constitution and the Scriptures, and hold others accountable for doing the same. George Washington himself said that religion and morality were "indispensable supports" to political prosperity.[116] To that I would add a hearty "Amen."

ABOUT THE AUTHOR

Tony Caruso is the son of Italian immigrants who came to America to build a better life among the coal mines and steel mills of western Pennsylvania. Thanks to a hard-working father who valued education, Tony earned a Bachelor of Science degree in engineering and has spent nearly thirty years in the profession. During his subsequent teaching tenure, he earned a Juris Doctor degree in law, and worked part time doing legal work for a Constitutional rights organization – something he intends to do full time in "retirement." He is passionate about school reform because of the excellent education he's received – one that was never available to his father, and is being denied to the current generation. He and his wife currently reside in Florida near their three children, and eight grandchildren.

ENDNOTES

1 National Center for Education Statistics, U. S. Department of Education, *Program for International Student Assessment*, 2006

[2] Chuck Noe, NewsMax.com, "Bush Decries Democrats' 'Soft Bigotry of Low Expectations'," Jan. 9, 2004

[3] U.S. Department of Education, National Center for Education Statistics, http://nces.ed.gov/programs/digest/d09/tables/dt09_144.asp

[4] Dr. W.R. Coulson, "The Role of Psychology in Current Educational Reform," Empire State Taskforce for Excellence in Educational Methods, 1997

[5] Dr. W.R. Coulson, *The Forerunner*, May, 1989

[6] Christine E. Johnson, "The A+ Plan and Student Achievement in Florida," Florida Institute of Education, June 2005

[7] Id.

[8] See www.answers.com/topic/employment-employers-perceptions-of-employment-readiness

[9] Robert Tai, "Block Scheduling: Not Helping High School Students Perform Better in College Science," June 2006, and www.johnwcooper.com/papers/blockscheduling.htm

[10] Benjamin Franklin, *The Writings of Benjamin Franklin*, Jared Sparks, editor (Boston: Tappan, Whittemore and Mason, 1840), Vol. X, p. 297, April 17, 1787

[11] John Adams, *The Works of John Adams, Second President of the United States*, Charles Francis Adams, editor (Boston: Little, Brown, and Co. 1854), Vol. IX, p. 229, October 11, 1798

[12] Ed Barnes, "Teacher Sues Over Right to Flunk Her Students," FOXNews.com, May 7, 2010

[13] 1895 eighth grade final exam, Salina, Kansas, www.rootsweb.ancestry.com/~kssvgs/graduationexam.htm

[14] House of Representatives, State of Florida, Council for Lifelong Learning, September 2001

[15] Robert Rector, *The Heritage Foundation*, Research report, April 8, 2002, and *National Survey of Adolescents and Their Parents*, Prepared for U.S. Dept. of HHS, February 26, 2009

[16] David Kupelian, *WorldNetDaily.com*, "What's behind today's epidemic of teacher-student sex?," March 22, 2006

[17] Id.

[18] John Adams, *The Works of John Adams,* Second President of the United States, Charles Francis Adams, editor (Boston: Little, Brown, and Co. 1854), Vol. IX, p. 229, October 11, 1798

[19] Ellen McCarthy, *The Washington Post*, August 16, 2009

[20] William Mattox, *The Washington Post,* Op-ed Feb. 13, 1994, Family Research Council survey

[21] Christine Kim & Robert Rector, *The Heritage Foundation*, Research report, "Evidence on the Effectiveness of Abstinence Education: An Update," February 19, 2010

[22] Id.

[23] Americans for Divorce Reform, http://www.divorcereform.org/edu.html

[24] Id.

[25] Matthew Daniels and Dan Englund, "Statutory Rape," *Massachusetts Family Institute*, October 1998, and Patricia Edmonds, "Teen Pregnancy Revives Laws on Statutory Rape," *USA Today,* March 28, 1996

[26] "America's Children: Key National Indicators of Well-Being," Federal Interagency Forum on Child and Family Statistics, 1997

[27] William Beckman, Illinois Right to Life Committee, Open Letter to Arne Duncan re Comprehensive Sex Education, June 27, 2006, and Brian Clowes, *Lifeissues.net,* "Condoms, STD, Teenagers, and International Case Studies Showing Condom Ineffectiveness Against HIV/AIDS," 2006, and Jana Mazanee, *USA Today,* "Birth Rate Soars At Colorado School.", May 19, 1992

[28] Robert Rector, *National Review Online*, "Abstinence Education Effective; Comprehensive Sex Ed a Big Flop," February 1, 2010, and http://www.pregnantpause.org/youth/sexed.htm

[29] *National Abstinence Education Association,* "Parents Prefer Abstinence Education 2 to 1, Zogby Survey Shows Dramatic Shift in Attitudes Once Parents Understand Differences Between Abstinence and Comprehensive Sex Education," May 3, 2007

[30] Christine Kim & Robert Rector, *The Heritage Foundation*, Research report, "Evidence on the Effectiveness of Abstinence Education: An Update," February 19, 2010

[31] "Teen Pregnancy Prevention," *Hearing Before the Subcommittee on Human Resources of the Committee on Ways and Means, House of Representatives*, Emory University survey, November 15, 2001

[32] Frederica Mathewes-Green, "Matters of Opinion: Free Love Didn't Come Cheap," *Christianity Today*, October 6, 1997

[33] Dr. Manny Alvarez, FOXNews Health, May 11, 2010

[34] "Student Suspended For Drawing Gun," *KPHO.com*, Phoenix, Arizona, August 21, 2007

[35] "Pupil Suspended for Drawing Soldier," *CNS News*, March 27, 2001

[36] Ann Coulter, *Guilty: Liberal "Victims" and Their Assault on America*, Crown Publishing, 2008

[37] John R. Lott, Jr., "Letting Teachers Pack Guns Will Make America's Schools Safer," *Los Angeles Times*, July 13, 2003

[38] Dave Kopel, "Follow the Leader, Israel and Thailand set an example by arming teachers," *National Review*, September 2, 2004

[39] "Ft. Hood Victims – Soldiers or Sitting Ducks?," *New American*, November 12, 2009

[40] "Brady Rankings: More Gun Laws, More Violent Crime," *NRA-ILA*, February 1, 2008

[41] "Spare the Rod or Swing the Paddle: How to Punish Students," *ABC News*, July 23, 2008

[42] "Islam studies required in California district," W*orldNetDaily*.com, January 11, 2002

[43] Nic Fleming, "Teachers and Social Workers Top League of Stressful Careers," *Telegraph.co.uk*, January 15, 2005

[44] Robert Rosenthal & Lenore Jacobson, *Pygmalion in the classroom* (1992) Expanded edition, New York: Irvington

[45] Jason Song, "Firing Teachers Can be a Costly and Tortuous Task," *The Los Angeles Times*, May 3, 2009

[46] Jami Lund, "Taking Employee Wages to Hijack Elections," *Evergreen Freedom Foundation*, June 2000

[47] **Davenport v. Washington Education Association**, 551 U.S.177 (2007)

[48] Proverbs 3:5,6, *The Bible*, NASB

[49] U.S. Department of Education, National Center for Education Statistics, 2008

[50] Stacey Childress, Jeff DeSimon, Nicholas G. Rupp, *Public Education in New Orleans: Pursuing Systemic Change through Entrepreneurship,* Harvard Business School Publishing, 2010

[51] Michael O'Sullivan, "Waiting for Superman," *The Washington Post*, October 1, 2010

[52] Joel Klein, Chancellor – NYC Dept. of Education, "Waiting for the Teachers' Union," *The Huffington Post*, September 24, 2010

[53] Laurel Rosenhall and Diana Lambert, "California Race to the Top Bills Give Parents More Say in Schools," *The Sacramento Bee*, January 11, 2010

[54] David Salisbury, "School Choice: Learning from Other Countries," *Cato.org*, May 31, 2005

[55] Joan Biskupic, "Supreme Court Leaves in Place Arizona School Tax Break," USA Today, 4/5/11. **Arizona Christian School Tuition Organization v. Winn,** Docket No. 09-987.

[56] Christopher J. Klicka, "Homeschoooled Students Excel in College," *Home School LDA*, September 20, 2006

[57] Brian D. Ray, Ph.D, "Facts on Homeschoooling, and Worldwide Guide to Homeschooling," *National Home Education Research Institute*

[58] Chuck Morse, "How Communist is Public Education?," *FreeRepublic.com*, March 23, 2002

[59] Id.

[60] http://educate-yourself.org/cn/communistmanifestortenplanks12sep06.shtml

[61] "What's Wrong With Outcome-Based Education?," *Phyllis Schlafly Report*, May 1993

[62] Thomas C. Tobin, "Bush Declares A+ Plan Success," *St. Petersburg Times Online*, November 14, 2003

[63] "Value Free Education "Bankrupt" says educator," *The Forerunner*, May, 1989

[64] Marissa DeCuir, "Schools Playing Down Valedictorian Honors," *USAToday*, June 28, 2007

[65] Nile Gardiner, Ph.D, & Morgan Roach, "Barack Obama's Top 10 Apologies: How the President Has Humiliated a Superpower," *The Heritage Foundation*, June 2, 2009

[66] Eugene Volokh, "School Children Singing the Praises of President Obama"(Apparently as a Public School Class Project)," *The Huffington Post*, September 25, 2009

[67] Sean Hannity, "Teacher Slams Flag Art, Praises Obama Picture," *FOX News*, May 12, 2010

[68] George Kiriyama, "Students Kicked Off Campus for Wearing American Flag Tees," *NBCBayArea.com*, May 6, 2010

[69] Byron York, *Michelle Obama: "Don't Go into Corporate America"*, *National Review Online*, February 29, 2008

[70] Kimball Perry, "Lawsuit: CPS Pushing Democrats," *The Cincinnati Enquirer*, October 18, 2010

[71] Abraham Lincoln, http://quotes.liberty-tree.ca/

[72] History Standards, *National Center for History in the Schools*, Basic ed. 1996, http://nchs.ucla.edu/standards.html

[73] Robert C. Johnston & Karen Diegmueller, "Senate Approves Resolution Denouncing History Standards," *Education Week*, January 25, 1995

[74] Karen R. Effrem, MD, "Federal Control of Education On Steroids," *EducationNews.org*, June 2, 2010

[75] Thomas Jefferson, Author, Declaration of Independence, http://quotes.liberty-tree.ca/quote/thomas_jefferson_quote_c09c

[76] James Madison, Speech, Virginia State Convention, December 2, 1829

[77] James Madison, Father of the Constitution, "*A Memorial and Remonstrance*", 1785: Works 1:163

[78] Jacqueline L. Salmon, "Most Americans Believe in Higher Power, Poll Finds," *The Washington Post*, June 24, 2008

[79] Associated Press, "Survey: Most Doctors Believe in God, Afterlife," *MSNBC.com*, June 23, 2005

[80] James Perloff, *Tornado in a Junkyard*, Refuge Books, 1999

[81] Lee Strobel, *The Case for a Creator*, Zondervan, 2004

[82] Institute for Creation Research, http://www.icr.org

[83] Affidavit of David Barton, recognized expert authority in American history, **ACLU v. McCreary County**, 361 F.3d 928, 2004

[84] M. D. Aeschliman, Murderous Science, National Review, March 28, 2005

[85] David Barton, *America: To Pray? Or Not to Pray?*, Wallbuilder Press, 2002

[86] Real Clear Politics, Direction of Country Polls, 2010

[87] Affidavit of David Barton, recognized expert authority in American history, **ACLU v. McCreary County**, 361 F.3d 928, 2004

[88] Noah Webster, preface to his *American Dictionary of the English Language*, 1828

[89] Benjamin Rush, *Letters of Benjamin Rush*, L.H. Butterfield, editor, Princeton University Press, 1951

[90] William O. Douglas, writing for the court, **Zorach v. Clauson**, 343 U.S. 306 (1952)

[91] **Engel v. Vitale**, 370 U.S. 421 (1962)

[92] Id.

[93] David Barton, *America: To Pray? Or Not to Pray?*, Wallbuilder Press, 2002

[94] Associated Press, "Valedictorian Sues Nevada School for Cutting Off Speech," July 15, 2006

[95] Marjorie Mazel Hecht, "Bring Back DDT, and Science With It!," *21st Century Science and Technology Magazine*, Summer 2002, and Todd Seavy, "THE DDT BAN TURNS 30 — Millions Dead of Malaria Because of Ban, More Deaths Likely," *American Council on Science and Health*, June 1, 2002

[96] Henry Payne, "As With the Truth About Acid Rain, the MSM Wants to Bury Climaquiddick," *National Review Online,* November 25, 2009

[97] U.S. Senate Minority Report, "More Than 700 International Scientists Dissent Over Man-made Global Warming Claims," U.S. Senate Environment and Public Works Committee, December 11, 2008

[98] Father Jonathan, "Middle School Gives Students Birth Control Pill," *FOXNews.com*, July 29, 2010

[99] Dan Springer, "School Abortion," *FOXNews.com*, March 24, 2010

[100] Sabine Siebold, "Merkel Says German Multiculturalism Has Failed," *Reuters*, October 17, 2010, and Dale Hurd, "'Islamization' of Paris a Warning to the West," *CBN News*, September 26, 2010

[101] Associated Press, "Arizona Immigration Law Faces New Legal Fight," *CBS News.com*, May 18, 2010

[102] Jana Winter, "Los Angeles Students to Be Taught That Arizona Immigration Law Is Un-American," *FOXNews.com*, June 2, 2010

[103] Id.

[104] "Nine Members of a Militia Group Charged with Seditious Conspiracy and Related Offenses," Dept. of Justice Press Release, March 29, 2010

[105] United Nations Treaty Collection, http://treaties.un.org/Pages/Treaties.aspx?id=4&subid=A&lang=en October 19, 2010

[106] Karen R. Effrem, MD, "Federal Control of Education On Steroids," *EducationNews.org*, June 2, 2010

[107] Thomas Sowell, "Parents with Backbone," *Jewish World Review*, February 26, 2004

[108] Daniel Czitrom, "Texas School Board Whitewashes History," *CNN.com*, March 22, 2010

[109] Thomas Jefferson, March 4, 1801, Draft of First Inaugural Speech

[110] Associated Press, "Evolution Theory Stickers Taken Off Textbooks," May 24, 2005, and "Judge Rules Against 'Intelligent Design,'" December 20, 2005

[111] Dave Kopel, "Follow the Leader, Israel and Thailand Set an Example by Arming Teachers," *National Review*, September 2, 2004

[112] Affidavit of David Barton, recognized expert authority in American history, **ACLU v. McCreary County**, 361 F.3d 928, 2004

[113] David Barton, *America: To Pray? Or Not to Pray?*, Wallbuilder Press, 2002

[114] Quote generally attributed to Edmund Burke, British statesman, 1795

[115] Kimberly Schwandt, "Florida's Teacher Union Fight: The Next Wisconsin?", *FOXNews.com*, March 4, 2011

[116] George Washington, First President of the United States, Farewell Address, 1796

www.ingramcontent.com/pod-product-compliance
Lightning Source LLC
Chambersburg PA
CBHW060458090426
42735CB00011B/2031